Vintage Rolex® sports models

A complete visual reference
& unauthorized history
Martin Skeet & Nick Urul

4880 Lower Valley Road, Atglen, PA 19310 USA

Library of Congress Cataloging-in-Publication Date

Skeet, Martin.
   Vintage Rolex Sports Models: a complete visual reference & unauthorized history / Martin Skeet & Nick Urul.
      p. cm
ISBN: 0-7643-1496-3
   1.   Montres Rolex S.A. 2.   Wrist watches—Switzerland—History—20th century. 3. Wrist watches—Collectors and collecting.   I. Title: Complete visual reference and unauthorized history. II. Urul, Nick. III. Title.
   TS543.S9 S54 2002
681.1'14—dc21
                                        2001005070

Printed in China
ISBN: 0-7643-1496-3

None of the companies referred to in this book either authorized or furnished or approved any of the information contained therein.
This book is derived from the authors' independent research.

Authors' notes
This book is unauthorized and the authors have received no help in any way from Rolex SA or any of its associated companies or any serving employee.

The following are registered trade marks of the Rolex Watch Company Limited, Geneva, Switzerland:
Cosmograph, the coronet logo, Datejust, Explorer, GMT-Master, Jubilee, Milgauss, Oyster, Perpetual, Prince, Rolex, Sea-Dweller, Submariner, Turn-O-Graph, Twinlock, Triplock, and we acknowledge the intellectual and proprietary rights of the company and all their subsidiary companies and distributors.
All trademarks are used for information & identification purposes only and no endorsement is implied.

Published by Schiffer Publishing Ltd.
4880 Lower Valley Road,
Atglen, PA 19310.

Phone: (610) 593 1777
Fax: (610) 593 2002
E-mail: Schifferbk@aol.com

Please visit our web site catalog at:
www.schiffer books.com

We are always looking for people to write books on new and related subjects. If you have an idea for a book please contact us at the above address.

This book may be purchased from the publisher. Include $3.95 for shipping. Please try your bookstore first. You may write for a free catalog.

In Europe, Schiffer books are distributed by:

Bushwood Books
6 Marksbury Avenue
Kew gardens
Surrey
TW9 4JF
England

Phone: 44 (0) 20 8392 8585
Fax: 44 (0) 20 8392 9876
E-mail: Bushwd@aol.com
Free postage in the U.K, Europe; air mail at cost.

Contents

Foreword   5
Introduction   7

Chapter   1   Rolex Chronology   8
Chapter   2   The Toolwatch Concept   14
Chapter   3   Submariner   18
Chapter   4   Sea-Dweller   46
Chapter   5   GMT-Master   56
Chapter   6   Explorer   68
Chapter   7   Milgauss   88
Chapter   8   Turn-O-Graph   98
Chapter   9   Cosmograph   106
Chapter   10   Bracelets   118
Chapter   11   Boxes   122
Chapter   12   Paperwork   126
Chapter   13   Movements   128
Chapter   14   Production Dates   130
Chapter   15   Collecting   132
Chapter   16   Advertisements & Literature   136
Chapter   17   Watches Sold at Auction   180
Chapter   18   Price Guide   208

Index   214
Bibliography   216

For Alison
- Martin Skeet

For Albeni and Kaplan
- Nick Urul

Foreword

At 11.30am on May 29th 1953 my father, Tenzing Norgay, stood atop Mount Everest - the highest and, for him, the most sacred mountain in the world.

Alongside him stood Sir Edmund Hillary, who, with his camera captured my father's wonder and exhilaration at the accomplishment which made him one of the world's great climbers. I am proud that this image of my father has become one of the historic icons of exploration.

Time plays a vital role in mountaineering. The climber must closely monitor his upward progress to ensure he leaves enough time to descend safely, and in daylight. Reduced oxygen at high altitude can play tricks on the mind, so a climber's life may depend on something as simple as attention to his watch. On that momentous day my father wore a trusted Rolex, a gift from the Swiss climber Raymond Lambert. He later gave that treasured watch to my brother Norbu. When I climbed Mount Everest in 1996, I wore a Rolex which the Tibetan explorer Heinrich Harrer had given to my father, and which my father had given to me. That watch was undeniably the most important tool on my climb.

My father and Hillary climbed Mount Everest through determination, strength, courage and teamwork. They also knew that their success relied on being attentive to time. The equipment they used looks prehistoric to today's climbers, but the importance of a reliable watch has not changed in fifty years.

In memory of my father, for the Sherpa people to whom I am proud to belong, and for all men and women who risk much in the pursuit of exploration, I am very happy to introduce this book on vintage Rolex sports models.

Jamling Norgay
Author, *Touching My Father's Soul: A Sherpa's Journey To The Top of Everest*

5

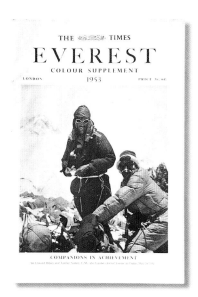

On the right, Sherpa Tenzing Norgay, and on the left, Sir Edmund Hillary, photographed after the first ascent of Mount Everest.

Acknowledgments

Many people gave time and energy towards
the creation of this book.

We would like to thank the irrepressible Sarah
Timewell, for her attention and dedication in
turning our rambling drafts into readable English
for the publisher; Edward Barber, for
photographing the watches; Lawrence
Houghton, for photographing brochures and
advertisements; Michael Harvey and Michael
Roberts, for photographing brochures and
scanning the auction catalog photographs;
Richard Chadwick, of Christie's UK and Jonathan
Darracott, of Sotheby's UK for lending us the
photographs reproduced on pages 183 to 205;
John Das, for the loan of the items shown on
pages 139, 140 and 141; Frank Nemeth, for the
loan of the items shown on pages 142, 143,
149, 174 and 175; Deborah Durban and Steven
Masefield, for research undertaken in Japan;
Peter Schiffer and Douglas Condon-Martin of
Schiffer Publishing.

All watches and other material shown is from the
authors' collections.

American editor
Douglas Congdon-Martin

English editor
Sarah Timewell

Introduction

This book began as a conversation in a bar. From the different perspectives of collector and dealer, we had each been looking at a vintage Rolex sports model we had never handled before. If only there were a book that showed all the Rolex sports models, we thought.

Previous books on Rolex watches did not cover the sports models in much detail, nor was the limited information they presented always accurate. In writing this book we have tried to rectify that gap.

As our research progressed we saw that there was no single, definitive history of any of the Rolex sports models. The history of a watch model could differ slightly from place to place, depending quite naturally on the specifics of the market or country in which it was sold. Consequently, in the interests of clarity and in writing this book for an international readership, we have chosen not to focus on any regional differences in how or when Rolex watches were named, released and sold. The release dates given are therefore the earliest known dates for a specific watch's release anywhere in the world.

Throughout this book the watch models are shown in the order of their release, so that the development of and gradual changes to the model will be clear. Diagrams have been used to illustrate subtle changes, such as alterations to the profile of the crown guards, and to the shape of the acrylic lenses, since these do not photograph very well.

The first Rolex sports models were usually manufactured in steel. Later, with the exception of the Explorer, Explorer II and Milgauss models, Rolex created bi-color and eighteen-carat gold versions of the sports models. The gold versions are referred to in the text, but because the sports models were created as working tools for specific professional or recreational use, we have shown only those made of steel, rather than the gold versions, which were intended for dress use.

Finally, we hope that the reader will find this book informative, helpful and enjoyable!

Martin Skeet and Nick Urul
London, May 2001.

7

Chapter 1

Rolex chronology

1881

Hans Wilsdorf, the founder of Rolex, is born in Bavaria, Germany, on March 22.

1900

Wilsdorf starts work with a watch and clock exporter in La Chaux de Fonds in Neuchatel, Switzerland.

1903

Wilsdorf moves to London, England.

WILSDORF & DAVIS

1905

With his business partner Alfred Davis, Wilsdorf founds Wilsdorf & Davis, in Hatton Garden, London. The company imports Swiss movements, sources dials and manufactures watches and watch-cases.

1907

Wilsdorf & Davis opens a technical office in La Chaux de Fonds, Switzerland.

THE ROLEX WATCH CO.

1908

The Rolex Watch Co. is registered as a company.

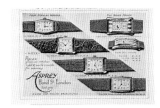

1910

A Rolex becomes the first wristwatch to obtain a First Class Chronometer Certificate from the Horology Society in Bienne, Switzerland.

1912

The Rolex Watch Co. begins manufacturing watches in Bienne, Switzerland.

1915

The company's name is officially changed from Wilsdorf & Davis to The Rolex Watch Co. Ltd.

1916

The company's technical office is moved from La Chaux de Fonds to Bienne, Switzerland.

1919

Wilsdorf returns to live in Switzerland, and in 1920 founds Montres Rolex in Geneva. Alfred Davis is now no longer involved in the company.

1925

The Rolex crown logo is registered as a trademark.

Rolex chronology

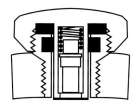

1926

Rolex patents the screw-down crown, known as the Oyster, it enables their watches to be completely waterproof.

1927

Wearing a Rolex watch, Mercedes Gleitze becomes the first British woman to swim the English Channel on October 21.

On November 27, the new Rolex Oyster is launched on front page newspaper advertisements, in the *London Daily Mail*.

1929

Rolex launches its new Prince watch, models 971, 971A, and 971U. The Prince becomes one of the 1930s most popular wristwatch designs.

1931

Rolex patents an automatic rotary winding mechanism, and incorporates it into the first Oyster Perpetual.

1935

Wearing a Rolex, Sir Malcolm Campbell sets the land speed record in his car *Bluebird*.

Five hundred Rolex Prince watches are manufactured for an order placed by the Royal household of King George V, to be given as gifts to mark the King's Silver Jubilee.

1937

The first Rolex watches marked Precision are released, models 3003 and 3004.

The first Oyster Zerograph is released, models 3642 and 3346, this watch features a calibrated bezel surrounding the dial, a feature that is the forerunner of the bezels used on many of the sports models of the 1950s.

1941

A new balance mechanism known as the Superbalance is developed and used in the Oyster and Oyster Perpetual 'bubbleback' models.

1945

The Oyster Perpetual Datejust is introduced - the first automatic, waterproof chronometer with the date visible through a window in the dial.

Late 1952

An Explorer prototype, model 6098, is launched. In May 1953, members of Sir John Hunt's Mount Everest expedition are issued with prototype Explorers.

Mid 1953

In September, a large test watch known as a Deep Sea Special is attached to the exterior of Professor Auguste Piccard's bathyscaphe, FNRS-2 which reaches a depth of 3131.80 meters (10,335 feet). The watch remains waterproof.

Late 1953

The Turn-O-Graph, model 6202, is introduced, available in steel, or steel and gold. Its distinctive case and bezel design will later be adapted to create the Submariner and GMT-Master.

Late 1953

The first Submariner is created. It is officially launched at the 1954 Basel Spring Watch Fair, in Switzerland.

Rolex chronology

Early 1954

The Milgauss, model 6541, is launched, having been designed for use in high magnetic fields.

Mid 1954

The GMT-Master, model 6542, is launched, which simultaneously displays the time in two different time zones.

The Oyster Perpetual Day-Date is also launched, it displays the date and day of the week through a window in the dial.

1959

The Submariner, model 5512, is launched, it introduces protective crown guards.

Early 1960

In January, a second bathyscaphe, the *Trieste*, reaches a new world record depth of 11,000 meters (35,798 feet) with Piccard, model 7205/0, attached to its exterior. The watch survives a pressure of over 1,086 atmospheres.

Mid 1960

Hans Wilsdorf, the founder of Rolex, dies in Geneva.

Late 1960

The first Cosmograph, model 6239, is launched. It features a tachymetric timing ring on the metal bezel, rather than on the dial. A few months later the 6241 model, with an acrylic bezel is launched.

1965

The Cosmograph, model 6240, which has screw-down waterproof pushers, is launched.

1966

The Sea-Dweller, model 1665, is created - its helium gas escape valve makes it the first commercially available watch for use by saturation divers.

1966

The Submariner, model 1680, which has a date feature, is launched.

1971

The Explorer II, model 1655, is launched.

1986

By 1986 more than half of all Swiss certified Chronometers produced have been manufactured by Rolex.

1990

Rolex produces its ten millionth Chronometer certified watch.

Chapter 2

The Toolwatch Concept

Today, watches like the Submariner, Turn-O-Graph, GMT-Master and Explorer are known as sports models. But when these models were conceived over forty years ago, they were among the first watches created for use in specific professions or recreational pursuits. This concept was known as the toolwatch, and although not a unique horological concept - in the 1920s Longines had manufactured a watch designed by Charles Lindbergh for pilots - it was a radical step away from the existing view, which was that, as required, ordinary watches would merely be adapted to suit specific uses. An example of this was Rolex's own First World War watch, which simply had a protective metal cage fitted over the glass.

The first sixty years of the Rolex Watch Company's history are a testament to how an organization can grow successfully through the use of constant innovation and product development. In the decade beginning in 1920, Rolex was

famous for its Prince models, and in the 1930s it was the bubble-back that other watch manufacturers copied. By the 1940s the durability and waterproofing of the bubble-back was being adopted across the entire Rolex range. This led the way, in the 1950s, for the development of the new models that are the sports models of today.

The tool watch owes its existence to Rolex's development of the first ever waterproof watchcase. After the company had produced this in the 1920s, continued development led to even tougher case design, and consequently the possibility of an increased depth rating. In the 1940s Rolex had made diving watches for the Panerai company which were issued to the Italian Navy. These were the first watches designed for use underwater and lessons learned in the construction were taken forward in the 1950s so that by the early 1950s, three watches existed which would form the basis of all

The Explorer, model, 6150 which was created in 1953.

The Turn-O-Graph, model 6202, which was created in 1953.

Oyster designs to follow - the 6150 Explorer, the 6200 Submariner and 6202 Turn-O-Graph. These were the prototypes of the tool watch, and they played a pivotal role in the subsequent history of all Rolex's sports models.

The Explorer should be considered a tool watch, because it was designed for extremely rugged use as well as absolute accuracy. Its toughness and reliability was ideal for the sort of adverse conditions experienced by mountaineers such as those from Sir John Hunt's Everest expedition, who successfully ascended the mountain in May 1953.

By 1953, the growth of Scuba diving pointed to a market for a specific diving watch. Jacques Cousteau was then formulating his theory that man would go to ever greater depths of the sea, and perhaps even live underwater for long periods of time. Scuba diving was as exciting a sphere of interest as space travel was to become in the next decade, and it had many more fans than actual divers. The Submariner was launched in 1953 and was an immediate success. The watch sold in large numbers - not only to divers but also (as it still does) to businessmen, who weren't averse to being associated with the daring new frontier of the deep sea.

It is important to remember how radical the Submariner was in the early 1950s. At that time the deepest scuba dive undertaken was to 90 meters (300 feet) - the fact that Rolex had created a watch with a depth rating of 180 meters (600 feet) suggested that even greater achievements in diving would come. At the same time the company positioned itself as a leader in the anticipation and development not only of divers' watches, but, as with the Explorer, of science and exploration.

Launched in 1953, the Turn-O-Graph had a waterproof case and a bezel that could be turned to read elapsed time. In its advertising it was promoted as the watch for the airline traveller; that it was

The Submariner, model, 6200, which was created in 1953.

The Toolwatch Concept

waterproof also indicates it was intended for the active customer. In 1966 Rolex was back at work developing an advanced watch for the diving company Comex (Compagnie Maritime d'Expertise). Comex's commercial saturation divers spent long periods in pressurised chambers breathing helium- and oxygen gasses. During the time spent in the chamber, the fine helium molecules were able to penetrate the acrylic crystal and gaskets of the watch. At the end of the work period, the divers were brought back to atmospheric pressure faster than the helium could escape the watch, causing a build-up of pressure that eventually pushed the crystal off the watch, and rendered it useless. To solve this expensive problem, Rolex fitted an ingenious gas escape valve to the Submariner and this became the basis for the Submariner Sea-Dweller 2000, released commercially in 1967 and rated to the impressive depth of 610 meters (2000 feet). Like the

Submariner, the Sea-Dweller is evidence of the commercial viability of the tool watch and continues to appeal to people who will never experience a saturation dive. It is also popular in landlocked countries!

In 1953 Rolex was asked by Pam Am to create a watch for airline crews that could display the time in the home country as well as the present location. The Submariner was adapted by adding an extra hour hand and a date feature, and named the GMT-Master. This watch was an instant commercial success for Rolex, and over forty years later it is still a major seller.

In the early-1950s Rolex began work on the watch that was to become known as the Milgauss. Workers in power plants and scientific researchers needed a watch that could operate accurately in high magnetic fields, so after adapting the tough Submariner case, Rolex used anti-magnetic alloys in the watch movement and placed it in an iron shield.

The Sea-Dweller, model 1665, which was created in 1966. It had a gas escape valve which enabled it to be used in saturation diving.

The GMT-Master, model 6452, which was created in 1954 for Pan American airline staff.

The Milgauss, model 6541, which was created in 1954. It was one of the worlds first truly anti-magnetic wristwatches.

16

The first Milgauss model looked very like the Submariner, and had the same bezel, but by the third model it also had a unique second hand not seen on any other watch. Ingenious though it was, it was not a big seller, and was later available only as a special order until it was discontinued in 1988.

In 1960 Rolex created the Cosmograph, a sophisticated chronograph which was marketed to the motor racing industry. It was ideally suited both to reading lap times to within a fifth of a second, and for calculating average speeds.

In the early 1970s Rolex created the Explorer II, which is perhaps the most unusual tool watch. Similar to the GMT-Master, it was designed for speleologists, who from the depths of their caves, would need to distinguish night from day. It had an extra hour hand, which rotated once every twenty-four hours, and was read against a fixed bezel to determine the hours of daylight and darkness (see illustration below). The global need for

such a watch is perhaps in doubt, but the Explorer II is still in the Rolex catalogues. One might say that it is indisputable evidence of the Rolex devotion to the tool watch concept.

Rolex's fifty-year commitment to the tool watch has resulted in some of the most enduring and appealing wristwatch designs produced by any manufacturer. Today the sports watches of every major company derive much of their styling and features from Rolex's groundbreaking designs, and its faith in the tool watch. Through the tool watch, Rolex became allied not just with dependability and accuracy, but with great human achievement.

The Cosmograph, model 6239, which was created in 1960.

The Explorer II, model 1655, which was created in 1971. This model had an extra hour hand, which can be seen pointing to eight o'clock.

Chapter 3

Submariner

Created in 1953, the Submariner was the world's first wristwatch designed for scuba diving.

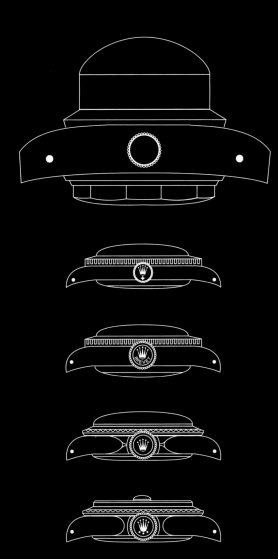

Submariner

The early 1950s saw the growth of the new sport of scuba diving, and Rolex was quick to recognise the importance to divers of having an accurate way of measuring time spent underwater, so they could avoid DCI (decompression illness), commonly known as the 'bends'. As a result, by 1954 there were three models of the Submariner, a new watch designed by Rolex for specialist underwater use.

The Submariner 6204 was officially launched at the Basel Spring Watch Fair in 1954. Fitted with the A260 movement, its dial did not display a depth rating, but in the accompanying catalog, Rolex certified the watch as waterproof to 180 meters (600 feet). Soon after, it was depth-rated to 200 meters (660 feet). Launched in the same year, the Submariner 6205 was depth-rated to 100 meters (330 feet), and used the same movement and watchcase as the 6204. The 6200 had a thicker case than the 6204 and 6205 models, and a larger winding crown marked *Brevet* (from the French word *brevette*, meaning patented). The 6200 was depth-rated to 200 meters and fitted with the A296 movement.

All three models had pencil-shaped, luminous hands, with a circle on the tip of the seconds hand. The bezel was divided into five-minute sections and was marked alternately with rectangles and numerals. None of the models had crown guards protecting the winding crown - these were not introduced until 1959, starting with the 5512 model. These early versions of the 6200, 6204 and 6205 usually did not have the word *Submariner* on the dial, which will be discussed on the following page.

Some dealers and collectors have assumed that the official 1954 launch in Basel was proof that the 6204 model was the first Submariner. Consequently it is also often assumed that the 6205 and 6200 were released after the 6204. However, careful examination of the evidence suggests that the 6204 was not the first Submariner.

First, it should be noted that Rolex used chronological numbering sequences for watch models: higher numbers signifed later launch dates. This would suggest that the first three Submariner models were the 6200, 6204 and 6205, in that order. Secondly, the A260 movement, fitted to the 6204 and 6205, in fact followed and improved upon the A296 movement, which was fitted to the

The Submariner, model 6200, launched in 1953.

6200. (Unlike watch model numbers, Rolex's movement numbers did not always conform to a chronological pattern.) Thirdly, it is important to examine the dials of these three Submariner models. Early versions of the 6200 model were fitted with the same type of dial used in 1953 on the Explorer models 6298 and 6150. At the twelve o'clock position on these dials is an equilateral triangle markedly different in shape from the triangles on the 6204 and 6205's dials (compare the 6200 and 6204 models shown on pages 26 and 27). This evidence suggests that in 1953 Rolex was already producing this watch for scuba diving, but that it was as yet unnamed a Submariner (just as the first Explorer-type watches existed before being given a model name). The 1954 Basel Spring Watch Fair was therefore an opportune moment to capitalise on the increasing success of scuba diving, and officially 'launch' the Rolex diving watch. The 6204 was formally christened the Submariner, and given a depth rating of 180 meters (600 feet).

However very soon after this, it appears that Rolex was either unsure of, or unable to use, the Submariner name as it does not appear on all subsequent models. Support for this hypothesis is lent by the fact that in 1954 Rolex registered a number of other model names, including Frogman, Skin Diver, Dive-O-Graph and Deep Sea Diver; the Submariner name was not finally registered to Rolex until the 1960s.

Also in 1954, the 6205 model appeared, and the first version did not display the word *Submariner* either. Confusingly, there exists a number of 6204 models on which *Submariner* has been obscured by black paint. (In these examples the black band is always the same size and in the same place, suggesting that it was done at the time of manufacture). The reason for this is unclear, but by late 1954, the Submariner name was consistently in use and all three models featured it on the dial.

Lastly, a 1970s Rolex brochure introduces an unexpected and interesting twist to the question of the launch dates for the 6200, 6204 and 6205. The text states that the company launched the first Submariner watch in 1953, and that this watch was depth-rated to 100 meters (330 feet). This statement points to two conclusions. One is naturally that the 6205 model was the first Submariner, but the evidence in favour of the 6200

The Submariner, model 6204, launched in 1954.

Submariner

(especially the chronological numbering) tends to outweigh any argument for the 6205. The second possibility is that either the 6204 or the 6200 model initially existed with the smaller depth rating of 100 meters (330 feet), rather than the respective180 meters (600 feet) and 200 meters (660 feet) which Rolex's accompanying official catalogs stated. Overall, it appears the likeliest conclusion to be drawn from all the evidence is that the 6200 was the first Submariner-type watch, and that when the 6204 model was released it initially existed with a depth rating of 100 meters, before being increased to 180 meters (600 feet).

Aside from the interesting question of release order, it appears that Rolex used a single dial blank for all three models, onto which (when they were there) the model name was printed in gold and the relevant depth rating printed in white. (This same Submariner dial blank was probably used on the first Milgauss watch, model 6541, which would explain why this rare, early Milgauss model has its name overprinted just above the six o'clock position, instead of below the twelve o'clock position, where it was later always placed.)

Like all Rolex watches, the Submariner was constantly being improved, and so

by late 1955 the model numbers were changed accordingly. The 6204 and 6205 models were both fitted with the new 1030 movement. The 6204 was re-numbered as the 6538 model, and the 6205 became the 6536 model. In addition to the new movement, the watches also differed from the previous models by the slightly larger winding crown. The 6536 model was later available with a chronometer version of the 1030 movement, as the 6536/1 model.

By 1956, the design of the Submariner hands changed - the luminous circle on the seconds hand moved further down the shaft, and the so-called Mercedes hands were adopted. The bezel also changed, and now had small rectangular markers showing the first fifteen minutes.

In late 1956 the 6538 model's watchcase was changed to the same thicker design as the 6200. To avoid confusion, 6538s now fitted with the thicker case carried the number 6538A. Once Rolex had used up the thinner cases, the 6538A model number reverted to 6538. At this time a new bezel was introduced for both the 6536 and 6538 models. The triangle at the twelve o'clock position was painted red, and the individual minute marks on the

The Submariner, model 6536/1. This model introduced a new bezel which had small rectangular markers showing the first fifteen minutes.

The Submariner, model 6538.

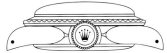

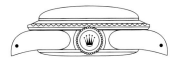

From top to bottom, the three versions of the crown guards, the square ended profile, the pointed and the rounded profile.

bezel were about half the length of those on the previous bezel. During this period the seconds hand on the Submariner (and on the Explorer and Turn-O-Graph) was painted white, and had a larger size luminous circle than before.

In early 1958 the 6200 was fitted with the 1530 movement, and renumbered 5510. By late 1958 the 6536/1 also had the 1530 movement, and was renumbered 5508. This model introduced a new square-shaped typeface for the numerals on the bezel, replacing the previously more rounded typeface. As an option, an Explorer-type dial with numerals at the three, six and nine o'clock position had been available on the 6538A, 6538 and 6200 models, and was also now offered on the 5510 model.

Today all the early Submariners without crown guards are known to collectors as James Bond models, after the Submariner's appearance in the first four James Bond films - despite the fact that it is only the 6538, 6200 and 5510 models that appear in the films. These days, Bond seems to have abandoned his previously impeccable taste for another watch manufacturer!

In 1959 the 5512 model was launched, powered by the 1570 movement, and

later the 1560. This model had the new design feature of protective crown guards. These first crown guards were square ended in profile, later a pointed profile was used, but in the mid-1960s they were replaced with a more rounded profile, when the Submariner's case became slightly thicker than before. The 5512, like the 6538 model, was also marked as an *Officially Certified Chronometer*. At this stage Rolex changed the wording on the chronometer watches to *Superlative Chronometer Officially Certified*. This was not at first marked on the dial. By 1965 the 6536 and 6538 models had ceased production, while the 5508 model was still being made.

In late 1962 the 5513 model made its debut, it was initially fitted with the non-chronometer 1530 movement. This new model also featured the pointed crown guards. Later it used the 1520 movement, with either 17 or 26 jewels, depending on the country in which it was sold.

Until the early 1960s both the 5512 and 5513 Submariners were produced with a gloss black dial. Until the early 1970s, both were available with the optional Explorer-type dial.

In 1966 a new Submariner, the 1680

The Submariner, model 5512. Launched in 1959, this model had the new design feature of protective crown guards.

The Submariner, model 1680. Launched in 1966, this model had a date feature.

Submariner

model, was created. This added a date feature to the Submariner range, and for the first few years the word *Submariner* was printed in red, before being gradually changed to white from 1974. The white printed dial was not launched into all the world markets at the same time, and dials which have the word *Submariner* printed in red are seen fitted to watches produced up to 1980. Both versions of the watch were fitted with the 1575 movement. The acrylic lens used on these models is very distinct, and stands almost 3 millimetres proud of the bezel. An eighteen-carat gold version of this watch was also available, and it is notable because it introduced gold surrounds to the luminous material on the dial, which were later adopted, in white gold, across the entire sports model range.

There has been a number of military adaptations of the Submariner, starting in 1955 when the 6204 and later the 6538 models were officially adopted by the British Royal Navy, and in 1956 by the Royal Canadian Navy. These watches had identifying service numbers engraved on the back, and the strap bars were soldered in place, so that they could only be fitted with a non-reflective cloth strap, rather than a metal bracelet. Apart from

these modifications the watches were exactly the same as those sold to the public. From the late 1960s, new British service procurement guidelines on wristwatch design led the British Royal Navy to commission a specially adapted 5513 Submariner with a high visibility hand design. In addition to the features on the 6204 and 6538 service models, the watchcase on this special 5513 model was satin-finished on the sides, to avoid the reflectivity produced by the usual mirror-finish. The dial had a 'T' in a circle, identifying the luminous material used as tritium. Another version, the 5517 model, was issued to the British Royal Marines. It had a unique bezel with sixty minute marks rather than only the first fifteen. During its issue the 5517 model was engraved with one of three forms of case marking; some had the case marked 5513 and had 5517 engraved on a case lug. Others had the case marked 5517, and the case back marked 5513. Some had both the case and case back marked 5517. Both the military 5513 and 5517 models were issued until the early 1980s when the contract with Rolex expired.

By the 1980s the Submariner's bezel had a new safety feature enabling it to only be turned anti-clockwise, rather than

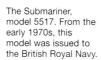

The Submariner, model 5517. From the early 1970s, this model was issued to the British Royal Navy.

The Deep Sea Special, created in 1953. Seven of these watches were created to test Rolex's waterproofing concept of screw-down crown, gaskets and tightly fitting acrylic crystal.

in both directions as before. Before this change, if the bezel was accidentally rotated clockwise while underwater, it could indicate that less dive time had elapsed than was the case. This could lead to the diver surfacing without carrying out the correct decompression stops, thereby running the risk of developing DCI (decompression illness).

Throughout the 1980s the Submariner range continued to be upgraded. In the early 1980s the 1680 date model was fitted with a scratchproof synthetic sapphire crystal, rather than the usual acrylic crystal, and its model number was changed to 16800. Shortly afterwards the dial itself was altered, with the luminous markings now being enclosed by white gold. The 5513 model was also fitted with the new dial design, and once the old cases with their acrylic crystals were used up, Rolex introduced a replacement model, which was also fitted with the synthetic sapphire crystal.

The oddest Submariner-type watches ever produced were the huge models named Deep Sea Specials and Piccard. Seven Deep Sea Specials were made in 1953 to test Rolex's waterproofing concept of screw-down crown, gaskets and tightly fitting acrylic crystal. Rolex

offered to supply the Swiss scientist, Professor Auguste Piccard, with watches for his planned dives in the bathyscaphe (depth-craft) *Trieste*.

The first experimental bathyscaphe *FNRS 2* had been designed in the late 1930s, but the outbreak of World War II had delayed its construction until 1947. In 1953 a second craft, *Trieste*, began a new series of dives. On 30 September 1953, with one of the specially constructed Deep Sea Specials fixed to the exterior of the bathyscaphe, and Auguste Piccard at the controls, the Trieste made a world record dive to 3155 meters (10,350 feet). On 23 January 1960, the *Trieste*, now piloted by Auguste's son Jacques, reached a new world record depth of 11,000 meters (35,798 feet), with a Piccard, model 7205/0, attached to its exterior, which survived a pressure of over 1086 atmospheres. Justifiably proud, Rolex still refers to this incredible feat of horological engineering in its promotional material. In June 2000 one of these rare Deep Sea Specials was sold at Christie's in London for £75,000/$112,500.

The Piccard, model 7205/0, created in 1958. One of these watches was fitted to the *Trieste,* and on 23 January 1960, reached a new world record depth of 11,000 meters (35,798 feet)

© Christie's

Shown at actual size

The enormous case of the Deep Sea Special. It differs from the later Piccard case by having a thicker crystal and a different winding crown design.

Submariner

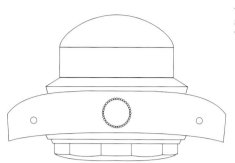

The enormous Deep Sea Special watchcase.

On the bezel fitted to the first 6200 and 6204 models the minute markers were thicker than those printed on the later 6205 bezel.

Deep Sea Special

Period
Early 1950s

Movement
1000/419343

Bracelet
40mm - rivetted construction

Seven of these watches were constructed for testing on Professor Auguste Piccard's bathyscaphe (depth craft).

On 30 September 1953 one of the watches, attached to the exterior of the craft, reached a depth of 3155 meters (10,335 feet). On surfacing it was intact and functioning correctly.

On 27 January 1960, a similar watch, the Piccard, reached a depth of 11,000 meters (35,798 feet) and suffered no damage.

Model 6200

Period
Early 1950s

Movement
A296

Bracelet
20mm - rivetted construction

It is most likely that the 6200 was the first Rolex diving watch, launched in 1953. The watch was depth-rated to 200 meters (660 feet). It had a thicker case than the 6204 and 6205 models, and had a larger winding crown marked *Brevet*.

Early versions did not have the word *Submariner* on the dial because a model name had not yet been created. Many of the first 6200 models have an Explorer-type dial.

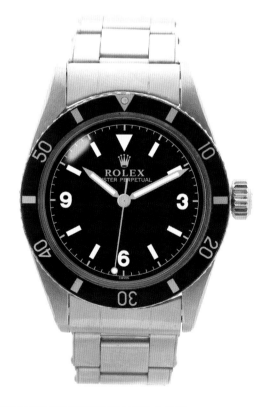

Key dates	1953	1953
	Deep Sea Special created	6200 model is launched

Model 6204

Period
Early 1950s

Movement
A260

Bracelet
20mm - rivetted
construction

The 6204 was the first diving watch to display the Submariner model name on the dial. The word *Submariner* was very small, and there was a large gap between the words *Oyster* and *Perpetual*.

It was probably released in 1953 and was officially launched at the Basel Spring Watch Fair in 1954. It may have originally been depth rated to only 100 meters (330 feet) but by 1954 was rated to 180 meters (600 feet).

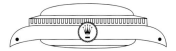

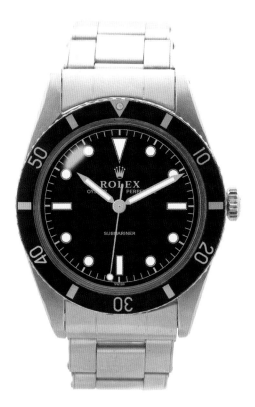

Model 6204

Period
Mid 1950s

Movement
A260

Bracelet
20mm - rivetted
construction

A number of 6204 models exist with the word *Submariner* obscured by black paint. The reason for this is unclear. On the dial of these (and later versions) there was a smaller gap between the words *Oyster* and *Perpetual*.

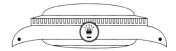

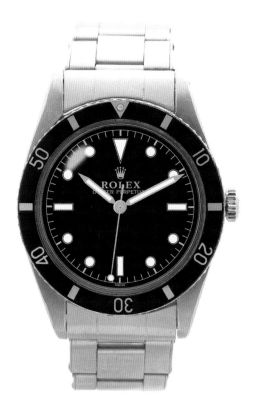

1953
6204 model is released, it is officially launched in 1954

Submariner

Model 6205

Period
Mid 1950s

Movement
A260

Bracelet
20mm - rivetted
construction

It is most likely that the 6205 was
the third Rolex diving watch,
launched in 1954.
    The watch did not always carry
a model name and was depth-
rated to 100 meters (330 feet).

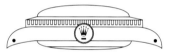

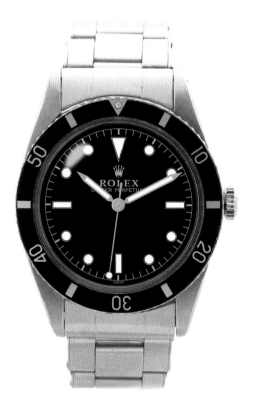

Model 6204

Period
Mid 1950s

Movement
A260

Bracelet
20mm - rivetted
construction

On the third version of the 6204
model the word *Submariner* was
printed on the dial and was larger
than on the first version.
    This model was issued to the
British Royal Navy.

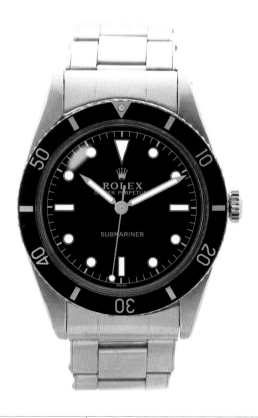

	1954	**1954**
	6205 model is launched	6204 model is issued to the British Royal Navy

Model 6538

Period
Mid 1950s

Movement
A260

Bracelet
20mm - rivetted construction

In 1955 the 6204 model was renumbered as the 6538 model. The newly-numbered watch used the same case as the 6204 model.

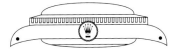

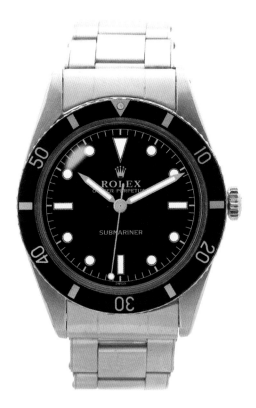

Model 6205

Period
Mid 1950s

Movement
A260

Bracelet
20mm - rivetted construction

In the mid 1950s, Rolex began to print the depth rating on the dial. The depth rating (printed in white paint) was added to an already printed dial on which the printing was in gold print.

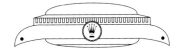

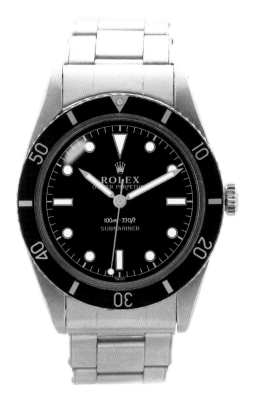

1955

6204 is renumbered as the 6538 model

Submariner

At this stage both the the triangle, and the luminous bubble on the bezel were larger than before.

Model 6536

Period
Mid 1950s

Movement
1030

Bracelet
20mm - rivetted construction

From 1955, the 6205 model was renumbered as the 6536 model.

This newly-numbered watch was powered by the 1030 movement and had a slightly larger winding crown than had been used on the 6205 model.

During this period the Mercedes hands were fitted to the Submariner range.

Model 6200

Period
Mid 1950s

Movement
A296

Bracelet
20mm - rivetted construction

Like the other Submariner models, the 6200 was now fitted with Mercedes hands.

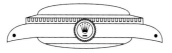

	1955
	6205 is renumbered as the 6536 model

The bezel now had small rectangular markers showing the first fifteen minutes.

On the bezel the rectangular minute marks were smaller than before and the triangle was now painted red.

Model 6536

Period
Mid 1950s

Movement
1030

Bracelet
20mm - rivetted
construction

The 6536 model sometimes had a depth rating printed on the dial in white.

Model 6536/1

Period
Mid 1950s

Movement
1030

Bracelet
20mm - rivetted
construction

The 6536 model was also available with a chronometer version of the 1030 movement. It was numbered 6536/1.

During this period the seconds hand was painted white.

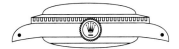

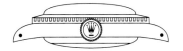

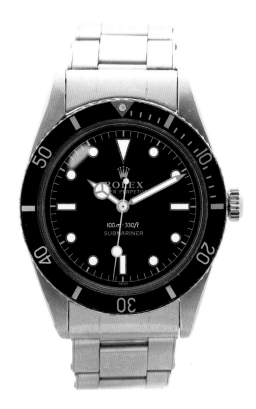

1956	1956
Minute marks added to the Submariner bezel	6536/1 model is launched

Submariner

Model 6538A

Period
Late 1950s

Movement
1030

Bracelet
20mm - rivetted
construction

In late 1956 the 6538 watchcase
was changed to the same thicker
design as the 6200 model.

Early versions were numbered
6538A. The watch was now
powered by the 1030 movement.

Model 6538

Period
Late 1950s

Movement
1030

Bracelet
Cloth strap

The 6538 model was officially
adopted by both the British
Royal Navy and the Royal
Canadian Navy.

The military watches had
identifying service numbers
engraved on the back, and the
strap bars were soldered in
place, so that they could only be
fitted with a non-reflective cloth
strap, rather than a metal
bracelet.

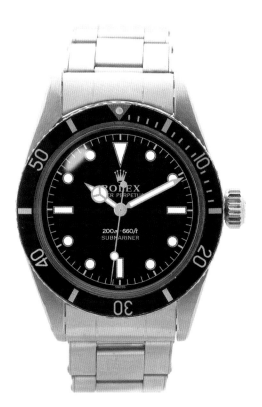

Late 1950s
6538 watchcase is changed to same thicker design as 6200 model

Model 6538

Period
Late 1950s

Movement
1030

Bracelet
20mm - rivetted
construction

On the second version of the
6538 model, the seconds hand
was painted white.

A depth rating and the words
Officially Certified Chronometer
were printed in white on the dial.

Model 6538

Period
Late 1950s

Movement
1030

Bracelet
20mm - rivetted
construction

The painted seconds hand was
not in production for long and
it was replaced by an
unpainted one.

Submariner

Model 6200

Period
Late 1950s

Movement
A296

Bracelet
20mm - rivetted
construction

The 6200 model was available
with a standard Submariner dial
or an Explorer-type dial.

Model 5510

Period
Late 1950s

Movement
1530

Bracelet
20mm - rivetted
construction

In early 1958 the 6200 model
was fitted with the 1530
movement, and was renumbered
5510.

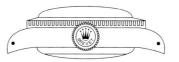

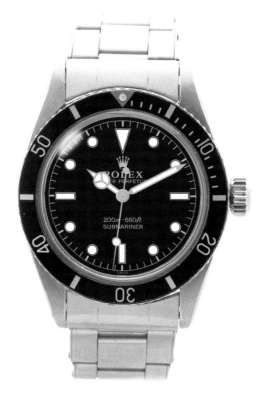

1958
5510 model is launched

This model introduced a new square-shaped typeface for the numerals on the bezel, replacing the round typeface. The luminous bubble on the bezel was now larger than before.

Model 5508

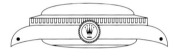

Period
Late 1950s

Movement
1530

Bracelet
20mm - rivetted construction

In 1958 the 6536/1 model was renumbered 5508.

The watch was powered by the 1530 movement.

Model 5508

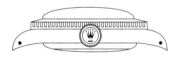

Period
Late 1950s

Movement
1530

Bracelet
20mm - rivetted construction

The 5508 model often had the depth rating printed on the dial in white.

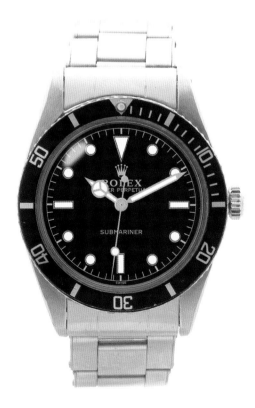

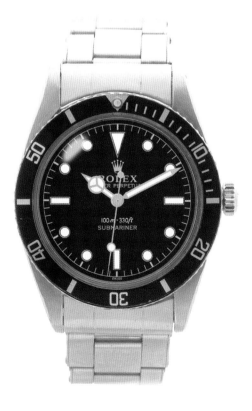

1958
5508 model is launched

Submariner

The 5512 model had a bezel with a larger bezel ring than was fitted to the 5508 model.

Model 5512

Period
Late 1950s

Movement
1570

Bracelet
20mm - rivetted construction

The 5512 model was launched in 1959. It had the new design feature of protective crown guards. The first versions of the crown guards were square ended in profile.

The 5512 model was not marked on the dial as a chronometer until the launch of the non-chronometer 5513 model.

Early versions were fitted with a gloss black dial.

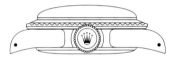

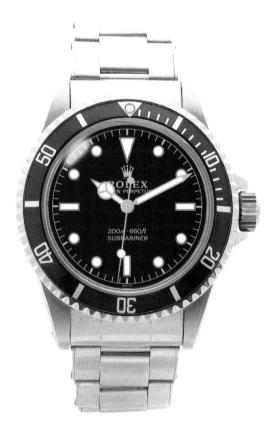

Model 5512

Period
Early 1960s

Movement
1560

Bracelet
20mm - rivetted construction

On early versions the depth-rating and the words *Officially Certified Chronometer* were printed on the dial in white.

From the early 1960s the watch was powered by the 1560 movement. The example has the second version of the crown guards which were pointed in profile.

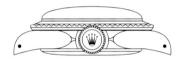

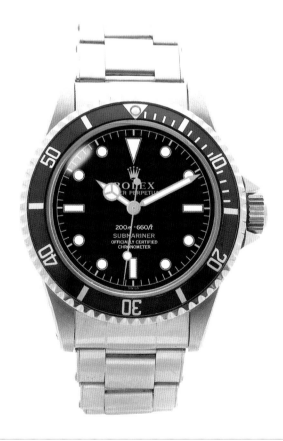

36

1959

5512 model is launched

Model 5513

Period
Early 1960s

Movement
1530

Bracelet
20mm - rivetted
construction

The 5513 model was launched in 1962. It was a non-chronometer rated watch, and used the same dial as the early 5512 model.

The watch was powered by the 1530 movement.

Early versions were fitted with a gloss black dial.

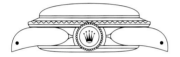

Model 5513

Period
Early 1960s

Movement
1530

Bracelet
20mm - rivetted
construction

Until the late 1970s, both the 5512 and 5513 Submariners were available with an optional Explorer-type dial.

At this stage the minute marks were positioned at the edge of the dial, rather than on a printed minute track inset on the dial.

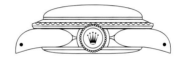

1962
5513 model is launched

Submariner

Model 5513

Period
Early 1960s

Movement
1530

Bracelet
20mm - rivetted
construction

During this period Rolex began to
change the wording at the bottom of
the dial from *Swiss* to *Swiss T<25*.
This was a gradual change and on
some watches *T<25* was added to
an existing dial on either side of the
word *Swiss*. These dials also often
had a white line printed below the
word *Submariner*. This was a
registration mark that ensured that
T<25 was correctly positioned when
it was added to the dial.

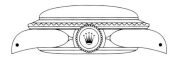

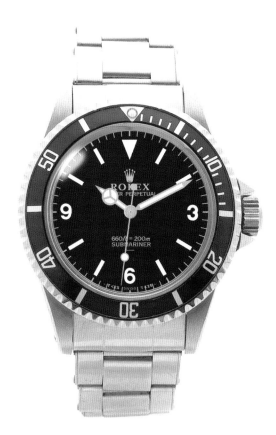

Model 5512

Period
Mid 1960s

Movement
1560

Bracelet
20mm - rivetted
construction

During this period Rolex changed
the wording on the chronometer
watches to *Superlative
Chronometer Officially Certified*.
This example is fitted with the
third style of crown guards which
had a rounded profile.

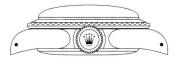

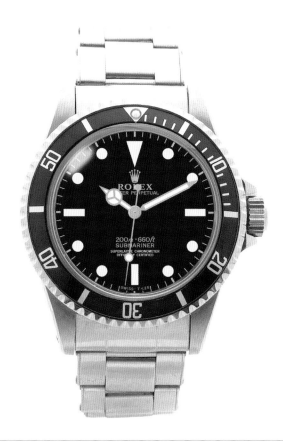

Model 5513

Period
Mid 1960s

Movement
1520

Bracelet
20mm - rivetted
construction

On the second version of the
5513 model the wording at the
bottom of the dial reads *Swiss
T<25.*

Model 5508

Period
Mid 1960s

Movement
1530

Bracelet
20mm - rivetted
construction

The 5508 model continued in
production until the mid-1960s.
The color of the dial printing
changed from gold to white.
  The word *Submariner* was
either positioned above or below
the depth rating on the dial.

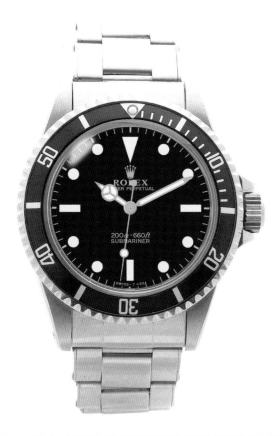

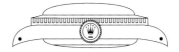

Submariner

Model 5508

Period
Mid 1960s

Movement
1530

Bracelet
20mm -rivetted
construction

During this period the dial on the 5508 model changed. The minute marks were positioned at the edge of the dial, rather than on a printed minute track inset on the dial.

Model 5510

Period
Mid 1960s

Movement
1530

Bracelet
20mm - rivetted
construction

The 5510 model continued in production until the mid 1960s. The color of the dial printing changed from gold to white.

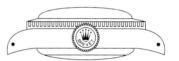

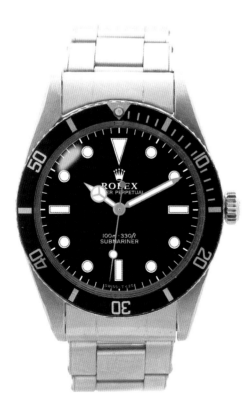

Model 5508

Period
Mid 1960s

Movement
1530

Bracelet
20mm - rivetted
construction

On the dial of the final version of
the 5508 model, the luminous
markings were not enclosed by a
printed border.

Model 5512

Period
Late 1960s

Movement
1560

Bracelet
20mm - rivetted
construction

In the late 1960s the color of the
dial printing used on the 5512
model changed from gold to
white. The model ceased
production in the early 1970s.

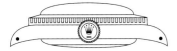

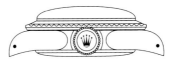

1965

6536 and 6538 models cease production

Submariner

Model 1680

Period
Late 1960s

Movement
1575

Bracelet
20mm - rivetted
construction

The 1680 model was launched in 1966. This added a date feature to the Submariner range, and for the first few years the word *Submariner* was printed in red. From 1974 *Submariner* was gradually changed to white. The white printed dial was not launched into all the world markets at the same time, and dials which had the word Submariner printed in red were fitted to watches produced up to 1980.

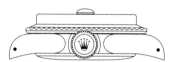

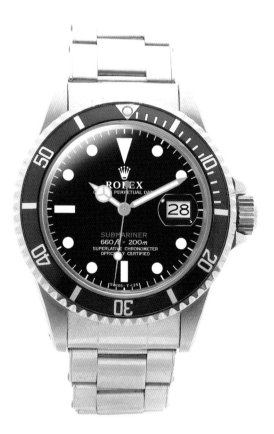

Model 5513

Period
Late 1960s

Movement
1520

Bracelet
Cloth strap

Specially adapted 5513 models were issued to the British Royal Navy. These had a high visibility hand design, and service numbers engraved on the back. On the case the strap bars were soldered in place, so that they could only be fitted with a non-reflective cloth strap. The case sides were satin-finished and the dial had a 'T' in a circle, identifying the luminous material as tritium.

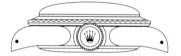

	1966		1968
	1680 model is launched		5513 model is adopted by the British Royal Navy

The 5517 model had a unique bezel with sixty minute marks rather than only the first fifteen.

Model 5517

Period
Early 1970s

Movement
1520

Bracelet
Cloth strap

In the early 1970s the 5517 model was issued to the British Royal Marines.

It had a unique bezel with sixty minute marks rather than only the first fifteen.

Model 5513

Period
Early 1970s

Movement
1520

Bracelet
20mm - rivetted construction

Throughout the 1970s an all-white print version of the Explorer-type dial was available on the 5512 and 5513 models.

1971

5517 model is created

Submariner

Model 5513

Period
Early 1970s

Movement
1520

Bracelet
20mm - rivetted
construction

From the early 1970s the minute
marks on the 5513 dial were
longer in length than those
previously used.

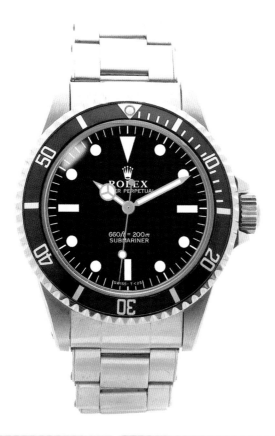

Model 1680

Period
Mid 1970s

Movement
1575

Bracelet
20mm - folded
construction

From 1974 the 1680 model was
available with all dial text printed
in white paint.

In some world markets the
dial was availble with the word
Submariner printed in red up
until 1980.

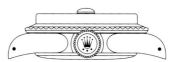

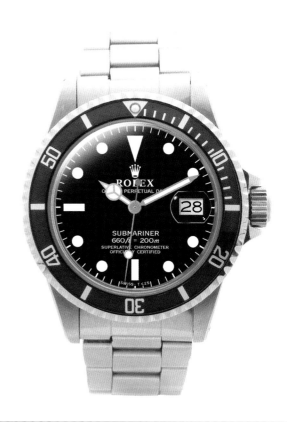

Model 16800

Period
Early 1980s

Movement
3085

Bracelet
20mm - solid
construction

In the early 1980s the 1680
model was renumbered 16800,
and was fitted with a synthetic-
sapphire crystal. It had a deeper
depth rating of 1000 feet (300
meters). Around 1984 the watch
received the new style dial which
had the luminous markings
enclosed by white gold.

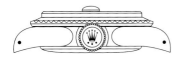

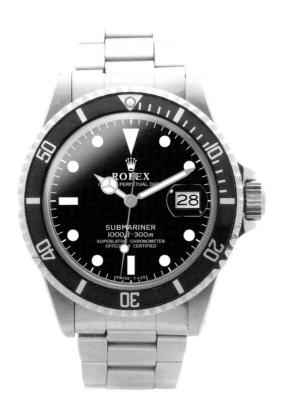

Early 1980s

1680 model is renumbered 16800

Chapter 4

Sea-Dweller   Launched in 1967, the Sea-Dweller was the
first wristwatch designed for use in
saturation diving.

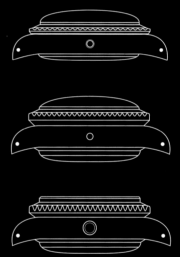

Sea-Dweller

The Sea-Dweller was created in response to a technical problem which divers for the French company Comex were experiencing with the Submariner 5513. Comex (Compagnie Maritime d'Expertise) contracted out commercial divers and equipment around the world, and was a leader in the development of diving technologies. In the 1960s its divers were using the conventional compressed air method as well as the new saturation diving technique. In compressed air diving, the deeper the diver goes, the longer he must spend slowly and safely decompressing on his way back to the surface. Decompression time is time spent not working, and with divers operating at ever-greater depths on oil and gas rigs, the technique of saturation diving was developed to maximise productivity while maintaining their safety.

It was known that underwater the human body absorbed into its tissues whatever gases it was breathing, until saturation was reached. The deeper the diver went, and the longer he spent underwater, the more gases he would absorb, and the more time his body would require to safely release those gases before he could resurface. The saturation diving technique replaced the use of compressed air, which is nitrogen and oxygen, with a helium-oxygen mix. The absence of nitrogen eliminated the threat of nitrogen-induced narcosis, which could be fatal to the diver, and at the same time, because helium escapes the body faster, less time was needed to decompress. It was also realised that if divers returned to the surface in a chamber pressurised to the depth at which they had been working, they could then remain – indeed live - in that chamber, until it was time to dive again. They would therefore only need to decompress once, at the end of the work period.

As saturation diving was more widely adopted, the problem with the Submariner 5513 began to occur. During the time spent breathing helium-oxygen in the pressurised chamber, the fine helium molecules were able to penetrate the acrylic crystal and gaskets

Comex issue
Submariner, model
5514.

Above, the prototype Submariner case for the Sea-Dweller, model 1665, used in 1966-67 which had a thicker bezel. Below, the thicker case introduced in 1967, which had a thinner bezel.

of the watch. At the end of the work period, the divers were brought back to atmospheric pressure faster than the helium could escape the watch, causing a build-up of pressure that eventually pushed the crystal off the watch, and rendered it useless.

Comex asked Rolex to try and find a solution, which the company rapidly did by installing a one-way pressure relief valve on the side of the watchcase, beside the nine o'clock position. This gas escape valve was patented in 1967, and is a masterpiece of engineering and simplicity (see diagrams on page 51). A small, spring-loaded piston was sealed to the exterior of the case with a waterproof O-ring, so that as the water pressure increased at depth, the piston was pushed more firmly against the O-ring, thus ensuring it remained waterproof. Later, during decompression, as the gas pressure inside the watch increased, the piston moved outward from inside the case, thus releasing the gas.

Rolex fitted a small edition of Submariner 5513s with the new valve for

Comex, the first models had standard Submariner dials, and a unique Comex identification number engraved on the back. Later 5513 models featured the Comex logo on the dial, as well as a Comex number on the back. In early 1967 when Comex requested a further batch, the watch was given an official model number, 5514. These 5514s were produced for the company in an edition of about 150 pieces. 5514 models featured the Comex logo on the dial, as well as a Comex number on the back.

In mid 1967 Rolex released a version of Comex's 5514 Submariner for sale to the public (this watch was created in 1966 but was not available to the public until 1967). The new watch, named the Sea-Dweller, model 1665, was fitted with the 1575 movement and given a thicker lens, which enabled it to withstand the staggering depth rating of 610 meters (2000 feet). This first Sea-Dweller released was a prototype model, it had a Submariner case which had a very high standing bezel surrounding the thicker lens. In late 1967 a thicker case was

49

Sea-Dweller, model 1665, launched in 1967.

Sea-Dweller

introduced. The back of the prototype 1665 watchcase was engraved with *Patent Pending* rather than *Patented* which is seen on later examples. The first Sea-dwellers are described on the dial in red print as *Submariner Sea-Dweller 2000*. This style existed for only six years before being replaced with an all-white print dial that carried only the Sea-Dweller name.

Note: There may have been an earlier 1665 model Sea Dweller that had the dial depth-rated to only 500 meters (1650 feet). It is unclear if this model was ever sold to the public, and we do not show it on the following pages.

Comex's saturation divers were now issued with 1665 white dial print Sea-Dwellers, which displayed the Comex logo on the dial, in place of the word *Sea-Dweller*. The first Comex model had a standard Sea-Dweller dial, and the Comex logo replaced the Sea-Dweller name. The second version of the Comex Sea-Dweller 1665 was marked on the upper part of the dial *Oyster Perpetual*, rather than *Oyster Perpetual Date*; and

on the lower part the depth rating was noted as 600 meters, rather than 610 meters. Comex's divers went on to achieve many world records, and in 1980 reached a new world record depth of 501 meters (1653 feet), undertaken whilst wearing their Sea-Dwellers. Comex and Rolex continued their association until the 1990s when Comex merged with another diving company.

In 1978 a new Sea-Dweller, model 16660, was launched. It was fitted with a synthetic sapphire crystal, and had an enormous depth rating of 1220 meters (4000 feet) as well as a larger, improved gas escape valve.

Today, the first 1665 Sea-Dwellers are highly sought after. The watch was not a big seller for Rolex, and since Sea-Dwellers were used by divers as working tools, few survive in good condition, with fewer still having any original paperwork. Until 1981, the original 1665 Sea-Dweller continued to appear alongside the 16660 model in Rolex catalogues.

Comex issue Sea-Dweller, model 1665.

Sea-Dweller, model 16660.

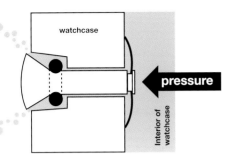
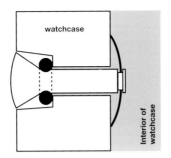

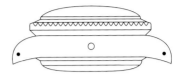

The first gas escape valve, created in 1967.

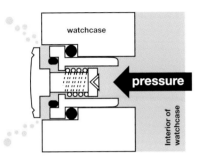
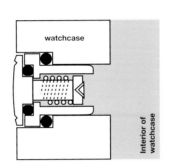

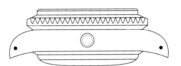

The second version of the gas escape valve, which was fitted to the 16660 model.

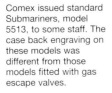

Comex issued standard Submariners, model 5513, to some staff. The case back engraving on these models was different from those models fitted with gas escape valves.

Left to right, case back engraving for a Comex issue 1665 from the mid 1970s and for the 16660 Sea-Dweller launched in 1978.

Sea-Dweller

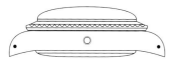

First version of the gas escape valve, which was fitted to the Submariner, model 5513.

First version of Comex case back engraving used on models fitted with the gas escape valve.

Model 5513

Period
Late 1960s

Movement
1520

Bracelet
20mm - rivetted construction

From 1967, specially adapted Submariner 5513 models were issued to Comex, a French commercial diving company.

These models had a one-way pressure relief valve installed on the side of the watch case which enabled them to be used in saturation diving.

A Comex identification number was engraved on the case back.

Model 5513

Period
Late 1960s

Movement
1520

Bracelet
20mm - rivetted construction

Not all the first versions of the Comex 5513 had a Comex logo printed on the dial.

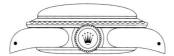

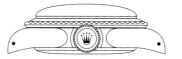

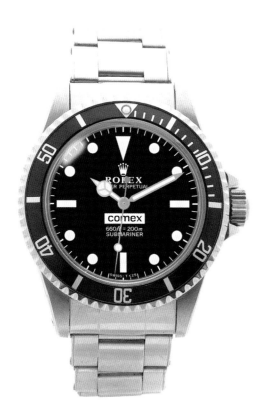

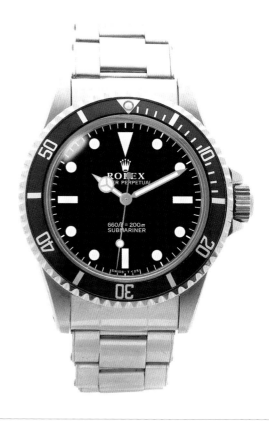

Key dates	**1967**
	Gas escape valve created

Case back for the 1665 Sea-Dweller. The first few watches produced were engraved with words *Patent Pending*, rather than *Rolex Patent*.

Model 5514

Period
Late 1960s

Movement
1520

Bracelet
20mm - rivetted construction

In1967, Comex ordered another batch of Submariners fitted with the one-way pressure relief valve, and the watch was given an official model number, 5514.

They were produced in an edition of about 150 pieces and had the Comex logo printed on the dial.

Model 1665

Period
Late 1960s

Movement
1575

Bracelet
20mm - rivetted construction

Launched in 1967, this watch was fitted with a thicker lens, enabling it to withstand the staggering depth rating of 610 meters (2000 feet). The first models released had a Submariner case, later a unique thicker case was used (shown below). The first Sea-Dwellers are described on the dial in red as the *Submariner Sea-Dweller 2000*.

53

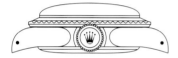

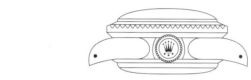

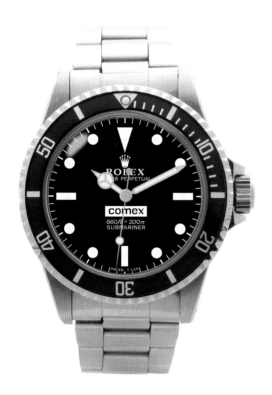

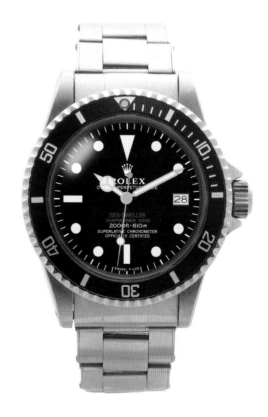

1967	**1967**
5514 model is launched	1665 model is launched

Sea-Dweller

Comex case back engraving for the 1665 model.

Model 1665

Period
Late 1960s

Movement
1575

Bracelet
20mm - rivetted construction

From 1967 Comex saturation divers were issued with 1665 Sea-Dwellers, which had an all-white print dial, and the Comex logo in place of the word *Sea-Dweller*. The first version was marked on the upper part of the dial *Oyster Perpetual*, rather than *Oyster Perpetual Date*; and on the lower part the depth rating was noted as 600 meters, rather than 610 meters.

Model 1665

Period
Early 1970s

Movement
1575

Bracelet
20mm - folded construction

The second version of the Comex Sea-Dweller had a standard Sea-Dweller dial, and the Sea-Dweller name was replaced by a Comex logo.

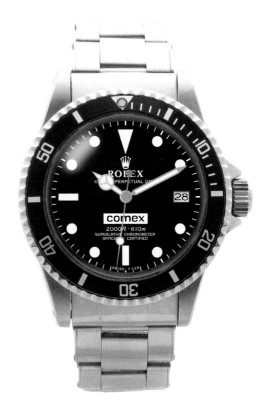

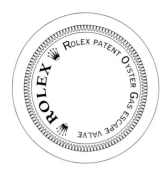

Case back engraving
used on 1665 models
from the mid 1970s.

Case back engraving
used on the 16660
model.

Model 1665

Period
Mid 1970s

Movement
1575

Bracelet
20mm - folded
construction

After 1974 the Sea-Dweller,
model 1665, was produced with
an all-white print dial. This model
did not have *Submariner 2000*
printed on the dial.

Model 16660

Period
Late 1970s

Movement
3035

Bracelet
20mm - solid
construction

In 1978 the 16660 model was
released. It was fitted with a
synthetic-sapphire crystal and
had a depth rating of 1220
meters (4000 feet). The watch
was fitted with a larger and
improved gas escape valve.

55

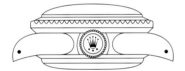

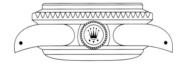

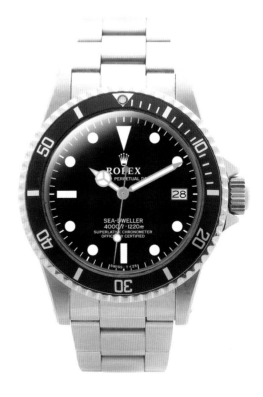

1978

16660 model is launched

Chapter 5

GMT-Master

Designed for Pan American Airline pilots and launched in 1954, the GMT-Master was the first watch to simultaneously display the exact time in two different time zones.

GMT-Master

In 1953 Rolex was approached by Pan American Airlines and a European airline (which could possibly have been the Italian carrier Alitalia), to develop a watch which could simultaneously display the exact time in two different time zones.

Pan Am and Rolex formed a working group, and plans were drawn up for a suitable watch. The model most appropriate for the necessary modification was a watch then under development and later launched as the 6204 Submariner. A date feature was added, and to help magnify the date a Cyclops lens could be fitted. Rolex brochures of the period assert that the Cyclops lens was an optional rather than a standard feature, until the early 1960s. However, since the watch is rarely seen without the Cyclops lens, it might be assumed that it usually left the factory with one fitted.

In 1954, after a year of development, the first GMT-Master, model 6542, was launched. This first model was very distinctive, and had the depth rating of 50 meters (165 feet) printed in red on the black dial. The watch had Mercedes hands. An extra, parallel-sided, hour hand had been added, which was clearly differentiated by the triangle at its tip. The hands went through very subtle changes over the years, mostly involving the shape and size of the triangle. An eighteen-carat

gold version was also available, and both versions were fitted with the 1065 movement (and later the 1036 and 1066 movements). On the first models, this triangle has a layer of luminous paint on top, whereas later models have a cutout within the triangle, which is filled with luminous paint. Until the early 1960s this triangle was about half the size of the later versions.

Instead of the Submariner's usual metal bezel insert, there was a special insert of transparent acrylic which had twenty-four hour markings printed on its underside. This material was chosen to cut down reflectivity and avoid dazzling pilots. The bezel ring in which the insert sat was also made of acrylic, and was coated in an alloy to resemble metal. It was found that this bezel ring had a degree of flexibility which could cause the bezel insert to fracture, and that, in very warm environments, the printing on the insert often flaked off. Rolex quickly replaced the acrylic insert with a metal one, and today GMT-Masters with their original acrylic bezel inserts command a high price.

As soon as it was released in 1954, Pan Am adopted the GMT-Master. To differentiate between watches intended for air crew and for ground staff, a special order with white dials was manufactured for ground staff. Probably fewer than two

The GMT-Master, model 6542, launched in 1954.

Pan American Airlines issue GMT-Master, model 6542, with a white dial.

hundred of these white dial watches were made, and they are a rare find today.

The GMT-Master was an immediate success, and was quickly adopted as the official timepiece of many of the world's leading airlines. A 1960 Rolex catalog stated that twenty out of twenty-one aircraft navigators considered it an important professional aid. (Cockpit crew in part determined their position by how much time had elapsed since take-off).

In late 1960, the second GMT-Master was launched, the 1675 model. The watch used the 1565 and later the 1575 movements, and it was fitted with protective crown guards. On the first models the profile of these crown guards is pointed, compared to the rounded shape later adopted. An eighteen-carat gold version of the watch was also available. The 6542 model was kept in production for a few years after the launch of the 1675 model, but after the mid-1960s was only available in eighteen-carat gold.

In the early 1960s the United States' space agency NASA decided to test a variety of chronograph watches for adoption as its official timepiece. The agency tested models from all the major manufacturers, including Rolex, before selecting Omega, who supplied NASA with the Omega Speedmaster that we today know as the Moon Watch, after Neil Armstrong wore it on the first moonwalk on July 21, 1969. Rolex may have missed out on a historic advertising coup, but in fact many of the Apollo astronauts purchased Rolex GMT-Masters in preference to the Omega Moon Watch, which despite its superior functions, must have been virtually impossible to operate when wearing bulky space-suit gloves. Rolex's own watch collection at its Geneva headquarters displays many GMT-Masters worn by Mercury and Apollo astronauts.

By the early-1970s, a Jubilee bracelet as well as the standard Oyster bracelet was also available on the 1675 model, making it the only sports model to be factory-fitted with this bracelet design. In the same period the bezel insert was also available in black. In 1981 the 1675 model was given the 3075 movement, and its model number was changed to 16750.

Four years later in 1985 the GMT-Master II, model 16760, was launched. This watch had a new dial with white gold surrounds to the luminous markers. The dial was now signed *Oyster Perpetual Date*. In 1986 this style of dial was also fitted to the 16750 model.

59

GMT-Master, model
1675, launched in 1960.

GMT-Master

The first bezel insert was manufactured of transparent acrylic which had the twenty-four hour markings printed on its underside. The bezel ring in which the insert sat was also made of acrylic, and was coated in an alloy to resemble metal.

Model 6542

Period
Mid 1950s

Movement
1036

Bracelet
20mm - rivetted construction

The 6542 model was the first GMT-Master, and was launched in 1954.

The first versions had a depth rating of 50 meters (165 feet) printed in red on the dial.

Model 6542

Period
Mid 1950s

Movement
1036

Bracelet
20mm - rivetted construction

The second version released did not display a depth rating on the dial.

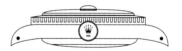

Key dates	**1954**
	6542 model is launched

Second version of
the bezel insert,
which was made of
metal.

Model 6542

Period
Mid 1950s

Movement
1065

Bracelet
20mm - rivetted
construction

The Cyclops lens, which
magnifies the date, was an
optional rather than a standard
feature until the early 1960s.

This example does not have a
Cyclops lens fitted.

Model 6542

Period
Mid 1950s

Movement
1066

Bracelet
20mm - rivetted
construction

On the third version launched, a
metal bezel insert replaced the
acrylic insert which had been
found to fracture.

In keeping with all sports
models of this period, the
seconds hand was painted white.

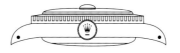

1955

Metal bezel insert introduced

Model 6542

Period
Mid 1950s

Movement
1066

Bracelet
20mm - rivetted construction

In 1954 Pan American Airlines adopted the GMT-Master as its official timepiece and issued it to staff.

A special batch with white dials was manufactured and issued to ground staff, in order differentiate between watches intended for air crew and ground staff.

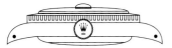

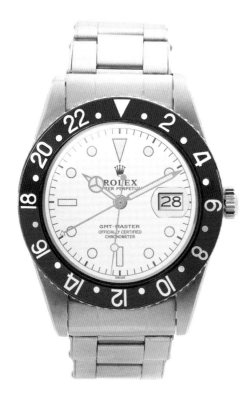

Model 6542

Period
Late 1950s

Movement
1066

Bracelet
20mm - rivetted construction

On the fourth version released the bezel ring in which the insert sat was made of metal.
At this time the seconds hand was no longer painted white.

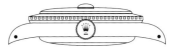

	1954
	Pan American 6542 model created

The bezel ring used on the 1675 model was larger than previously used.

Model 1675

Period
Early 1960s

Movement
1565

Bracelet
20mm - rivetted construction

The 1675 GMT-Master was launched in 1960.

This model introduced the protective crown guards to the GMT-Master range. On the first versions of the watch the profile of the crown guards was pointed.

On this example the crown guards are pointed, and the triangle at the tip of the secondary hour hand is curved where the long side joins the hand.

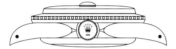

Model 1675

Period
Early 1960s

Movement
1565

Bracelet
20mm - rivetted construction

On the second version of the 1675 model the triangle at the tip of the secondary hour hand was almost twice as large as that used on the previous model.

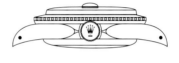

1960
1675 model is launched

Model 1675

Period
Early 1960s

Movement
1565

Bracelet
20mm - rivetted
construction

During the early 1960s, the color of the printed lettering on the GMT-Master was changed from gold to white.

This was carried out gradually and many watches are seen with both colours printed on the dial. Often, as here, the minute track is printed in gold whilst the rest of the printing is in white. At this stage the numerals printed on the date disk were black.

64

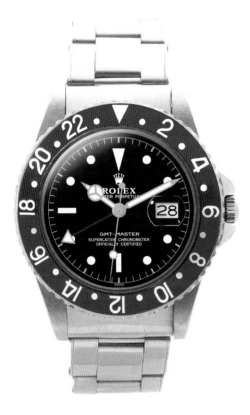

Model 1675

Period
Mid 1960s

Movement
1575

Bracelet
20mm - rivetted
construction

The fourth 1675 version had an all white print dial.

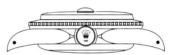

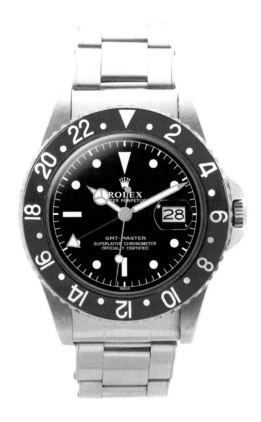

Model 6542

Period
Mid 1960s

Movement
1066

Bracelet
20mm - rivetted
construction

The 6542 model was not immediately replaced by the 1675 model, and later versions had an all white print dial.

At the bottom of the dial the wording was changed from *Swiss* to *Swiss T<25,* indicating that the luminous material had changed from luminova to tritium.

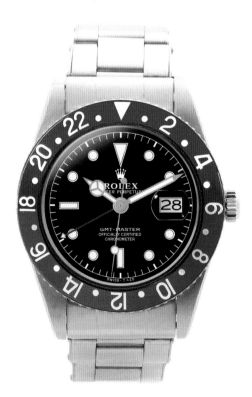

Model 1675

Period
Late 1960s

Movement
1575

Bracelet
20mm - rivetted
construction

In the late 1960s Rolex changed the design of the dial fitted to sports models. The new dials did not have printed borders enclosing the luminous paint.

This version has the new rounded crown guards which replaced the pointed versions. At this time the material used to manufacture the hands changed from gold colored metal to stainless steel.

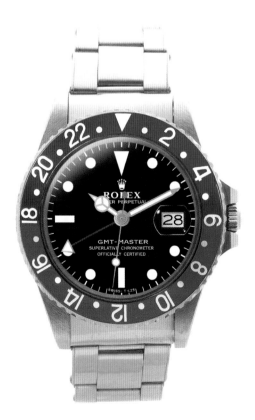

GMT-Master

From the early 1970s the bezel insert was also available in black.

Model 1675

Period
Late 1960s

Movement
1575

Bracelet
20mm - folded
construction

On the second version of the new dial the minute marks were larger in length than those previously used.

Model 1675

Period
Early 1970s

Movement
1575

Bracelet
20mm - solid
construction

By the early 1970s, a Jubilee bracelet as well as the standard Oyster bracelet was available on the 1675 model. It was the only sports model to be factory-fitted with this bracelet design.

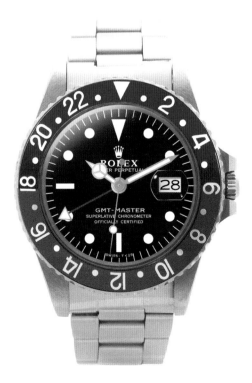

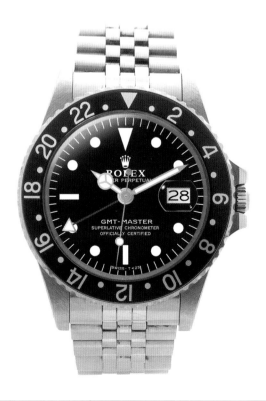

c.1974

Jubilee bracelet is available on the 1675 model

Model 16750

Period
Early 1980s

Movement
3075

Bracelet
20mm - solid
construction

In 1981 the 1675 model was
renumbered as the 16750 model,
the newly-numbered watch
looked identical to the 1675, but
was now fitted with the 3075
movement.

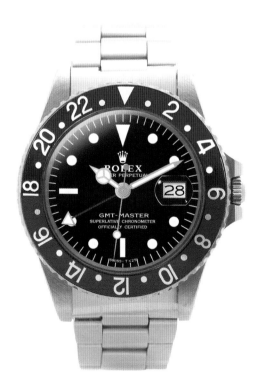

1981
16750 model launched

Chapter 6

Explorer

Created in 1953, the Explorer was intended for rugged expedition use. It was issued for use on many notable expeditions, including the one led by Sir John Hunt which successfully ascended Mount Everest in May 1953.

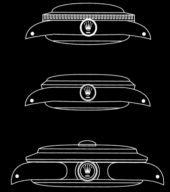

Explorer

The Explorer owes its existence to Rolex's efforts to continually improve its waterproof cases and shock-proofing systems.

By the early 1950s Rolex's reliable, rounded back watches, known as bubble-backs, had gained the reputation for being extremely tough and hard-wearing. It was therefore a logical step for Rolex to create a new model utilising the sturdy features of the bubble-back, and market it specifically to the more active and adventurous customer.

From late 1952, prototypes of the first Explorers were given to mountaineering and other expeditions for field trials. At this stage the Explorer name was not displayed on the dial, and the watches were simply bubble-backs with white dials, on which the three, six and nine were marked with steel or brass arrowhead symbols. The watches had leaf-shaped hands in a choice of steel or brass.

Several members of Sir John Hunt's May 1953 Everest expedition were issued with these watches, and it has long been assumed that Tenzing Norgay wore one for the successful summiting of Everest on 29 May. However, Tenzing's

son, Jamling Norgay, has confirmed to the authors that the watch worn by his father that day was a gold bubble-back given to him by the British climber Raymond Lambert. Hillary actually wore a Smiths watch for the climb, and fortunately for Rolex's publicity and marketing departments, he and Norgay refused to reveal which of them had actually set foot on the summit first. Both Smiths and Rolex reported the role of their watches in this great achievement, but it is the Rolex Explorer which is forever linked with the conquest of Everest. This impression was reinforced by the fact that Rolex had issued Hillary's 1952 Cho Oyu expedition with watches, and also went on to issue his later expeditions with Explorers, which was featured in the company's promotional material. In fact, Hillary revealed in the 1990s that he had been the first to reach the summit, so a Rolex was actually the second watch on the top of the world.

The original photographs of the 1953 Everest expedition, in the collection of the Royal Geographical Society in London, show that these Explorer prototypes had white dials which were not marked with the Explorer name. A Rolex

The Explorer prototype, model 6098, this watch was issued to Sir John Hunt's May 1953 Everest expedition.

The Explorer, model 6298, this watch was the first Explorer to have a recognisable Explorer-type 'Quarter Arabic' dial.

advertisement from 1953 (shown on page 140) showing the watch provided to the expedition appears to confirm this. A watch issued to expedition member Tom Bourdillon, sold at auction in the 1990s, had a recognisable Explorer-type dial which was black rather than white, and also did not display the Explorer name. This was most likely a 6150 model, so it appears there were two prototype Explorer watches issued to the Everest expedition.

The first two Explorer models (6098 and 6150) were launched in the early 1950s. By mid 1953 the 6098 had become the 6298 model, and by late 1953 the 6150 had become the 6350 model. There was a transition period during which these earlier model numbers, engraved on the inside of the watchcase, were struck through, and below the new numbers were engraved. All these models used the A296 movement. The 6298, 6150 and the 6350 had a recognisable Explorer-type 'Quarter Arabic' dial with the three, six and nine marked with luminous numerals on a black ground. Promotional material for the Explorer described it as the 'reinforced Oyster', and emphasized its

tough construction and reliability in harsh conditions. Like the 6202 Turn-O-Graph, the Explorer's dial was available with a textured finish, described by Rolex as a honeycomb dial. This was an attempt to soften its austere look and ensure a percentage of sales from the less adventurous customer.

The first four Explorers were followed in c.1955 by the 6610 model. Fitted with the 1030 movement, the 6610 initially had pencil-shaped hands similar in style to the early Submariners, and later the Mercedes hands were used. This was the last Explorer to be fitted with a honeycomb dial. The next Explorer, model 5504, launched in the late 1950s, had the 1530 movement. The early 5504 dials are finished in a very attractive gloss black, rather than the matt black which was later used, and are marked *Precision* or *Super Precision* on the dial.

By 1959 the 1016 model had appeared. This Explorer had a simple black dial, which was gloss black at first; later a matt black dial was fitted. It was initially fitted with the 1530 movement; later the 1560 movement was used. In the 1970s the 1570 calibre movement replaced the 1560, and later the watch

The Explorer, model 6610, this first version had a depth rating printed in red on the dial.

The Explorer, model 1016, launched in 1959.

Explorer

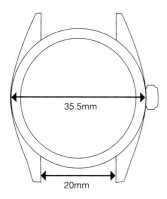

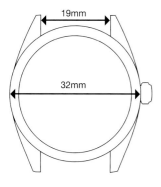

Top, the Explorer 1016 model watchcase, above, the Explorer 5500 model watchcase which is smaller and can only be fitted with a 19mm bracelet.

also came with a 'hacking' feature. This enabled the movement to be stopped exactly on the beat of a second, to allow accurate time-setting. This particular feature, and the 1016's successful blend of the sports and dress watch look, has made it very attractive to today's collectors. The 1016 model remained in the Rolex catalog until 1989, when it was removed prior to its replacement by the 14270 model.

From the late-1950s Rolex was enjoying a rapid worldwide growth in sales, especially in the North American and Asian markets. Eager to capitalise on the success of the Explorer, which had a distinctive dial, the company began to use the same dial design on the Air-King 5500 model. The Air-King was a dress watch with a smaller case than the Explorer, and a 19mm width bracelet rather than the 20mm bracelet used on other Explorer models. This watch used the 1520 movement, and later the 1530, and the dial was marked either *Precision* or *Super Precision*. On the early models a gloss black dial was fitted. These small Explorers were not popular with collectors until recently, in part because many people who were unfamiliar with

the model assumed they were fakes. However, current interest in all Explorers has meant that the price of the 5500 model is also rising. Unfortunately for the buyer, only Rolex's records would reveal if the watch left the factory with an original Explorer dial, or had a fake one fitted later.

Other Explorer variations include 6298 and 6299 models, many of which were sold to the Canadian market. These are just dress watches with white dials, and they often have bi-color cases. The Explorer date, model 5701, is another variation. It had a black dial and arrowhead markers. All of these watches look very different from what collectors would recognise as an Explorer, and for this reason they are not shown on the following pages. The vintage literature section of this book shows a 1962 Canadian catalog which pictures some of these models.

The Space-Dweller model is another variation of the Explorer. This model was launched to commemorate a 1963 visit to Japan by the astronauts from NASA's Mercury Space Program. The production run for these watches was extremely small, and they were not released into all

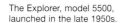

The Explorer, model 5500, launched in the late 1950s.

The Space-Dweller, model 1016, launched in 1963.

of Rolex's world markets.

In 1971 the Explorer II, model 1655, was introduced. It was fitted with either the 1570 or 1575 movement, depending on what was available in Rolex's stock. One of Rolex's more unusual creations, it testifies to the company's commitment to the 'tool watch' – namely, a watch designed for specific professional or recreational use. The Explorer II is simply an adaptation of the GMT-Master, with a fixed bezel. It was intended for use by speleologists, and had an extra hour hand reading against the fixed bezel, thus marking the hours of darkness or daylight. On the first models the seconds hand did not have a luminous circle, and the extra hour hand was orange. On models released after 1974, the luminous circle was added and on models released after 1975 the extra hour hand was painted red.

The 1655's design was strongly influenced by early 1970s style, and some might say that the dial appears cluttered and illegible. Like many other Rolex watches which did not enjoy strong sales during their main production years, the first Explorer II models are today popular with collectors - but they were difficult to sell in the late 1980s and early 1990s. One might wonder how much of this current popularity is based on the resurgence of 1970s style, rather than on a genuine estimation of the Explorer II as a design landmark.

In mid-1985 the 1655 model was removed from the Rolex catalog and replaced in late 1985 by the 16550 model. This new model looked very different from its predecessor. It was fitted with a synthetic-sapphire crystal, and had a new bezel design and 'Mercedes' hands. The watch was fitted with the 3085 movement, and for the first time the Explorer II came with a choice of a white or black dial. The white dial watches have become very collectible - the paint used was a pale ivory color, which faded to a distinctive cream that collectors have found especially attractive. These first models have white gold hands and white gold settings around the dial's luminous markings. By 1989 Rolex had changed the white dial paint to a non-fading bright white, and the bezel's numerals were represented with a new typeface. The settings around the luminous markings and the Mercedes hands were now painted black.

The Explorer II, model 1655, launched in 1971.

The Explorer II, model 16550, launched in 1985.

Explorer

Model 6098

Period
Early 1950s

Movement
A296

Bracelet
20mm - rivetted
construction

The 6098 model is a prototype of
the Explorer.

Several members of Sir John
Hunt's successful May 1953
Everest expedition wore these
watches.

This watch was a bubble-back
with a white dial and did not
display the name *Explorer* on the
dial.

Model 6298

Period
Early 1950s

Movement
A296

Bracelet
20mm - rivetted
construction

In 1953, the 6098 was
renumbered as the 6298 model.

This was the first model to
feature the Explorer-type Quarter
Arabic dial but did not have the
word *Explorer* on the dial, which
was marked *Precision* rather than
Officially Certified Chronometer.

It seems likely that this type of
dial was also fitted to some of
the watches issued to Sir John
Hunt's successful May 1953
Everest expedition.

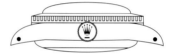

Key dates	**1953**	**1953**
	6098 model launched	6298 model launched

Model 6150

Period
Early 1950s

Movement
A296

Bracelet
20mm - rivetted
construction

The Explorer, model 6150, was
the first to be fitted with
Mercedes hands. The early
versions of these hands had a
longer hour hand than was later
used, and the minute hand had a
point at the tip.

Model 6350

Period
Early 1950s

Movement
A296

Bracelet
20mm - rivetted
construction

By late 1953 the 6150 was
renumbered as the 6350 model.
During this period the model
name Explorer was launched and
was printed on the watch dial.

Early 1953	**Late 1953**
6150 model launched	6350 model launched

Explorer

Model 6350

Period
Mid 1960s

Movement
A296

Bracelet
20mm - rivetted
construction

Many early Explorer models were
fitted with pencil-shaped hands of
the same style used on the
Submariner. In mid 1954 Rolex
launched flush-fit endpieces.
These were then fitted to the
sports models.

The dial on this example has a
textured finish, described by Rolex
as a honeycomb dial. This was
available on all of the first Explorer
models.

Model 6610

Period
Mid 1950s

Movement
1030

Bracelet
20mm - rivetted
construction

The Explorer, model 6610, was the
fourth Explorer and was launched
in 1955.

On the first versions, the depth
rating was printed in red and
Explorer was printed on the lower
half of the dial. Watches from this
period were marked *Officially
Certified Chronometer*.

On this example, the seconds
hand has the luminous circle
positioned further down the shaft.

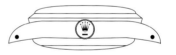

1955

6610 model launched

Model 6610

Period
Mid 1950s

Movement
1030

Bracelet
20mm - rivetted
construction

The second version of the 6610 model had *Explorer* printed on the upper half of the dial and a depth rating was printed below it.

At this stage all sports models had the Mercedes hands fitted.

Model 6610

Period
Mid 1950s

Movement
1030

Bracelet
20mm - rivetted
construction

On the third version, there was no depth rating printed on the dial. On all sports models from this period the seconds hand was painted white.

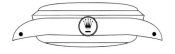

Explorer

Model 6610

Period
Late1950s

Movement
1030

Bracelet
20mm - rivetted
construction

The Explorer, model 6610, was
the last Explorer to be fitted with
the honeycomb dial.

Model 5500

Period
Late 1950s

Movement
1530

Bracelet
19mm - rivetted
construction

During this period Rolex began
to fit an Explorer-type dial to the
Air-King, model 5500.

The Air-King model had a
smaller case than the Explorer
and was fitted with a 19mm
width bracelet rather than the
20mm bracelet used on other
Explorer models. The dial was
marked either *Precision* or *Super
Precision* and the early versions
were in gloss black.

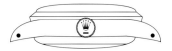

c.1957

5500 model launched

Model 5504

Period
Late 1950s

Movement
1530

Bracelet
20mm - rivetted
construction

The 5504 model used the same watchcase as the 6610. The dials were often marked in white print as *Super Precision,* and the minute marks were longer than those used on the 6610 model.

Early versions were fitted with a gloss black dial.

Model 1016

Period
Late 1950s

Movement
1530

Bracelet
20mm - rivetted
construction

The 1016 model was launched in 1959. On the dial the minute marks were positioned at the edge, rather than on a printed minute track inset on the dial.

It was the first Explorer model to be marked *Superlative Chronometer Officially Certified* on the dial and to have *Swiss T<25* at the bottom of the dial.

Early versions were fitted with a gloss black dial.

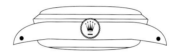

c1957	1959
5504 model launched	1016 model launched

Explorer

Model 5500

Period
Late 1950s

Movement
1530

Bracelet
19mm - rivetted
construction

During this period the 5500
model was produced with an all-
white print dial. The minute
marks were positioned on a
minute track which was inset on
the dial.

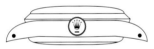

Model 5500

Period
Early 1960s

Movement
1520

Bracelet
19mm - rivetted
construction

On the third version of the 5500
model dial, the minute marks
were positioned at the edge of
the dial.

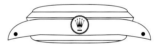

RLX

The slab serif typeface used to write *Rolex* on the dial of 1016 models from the early 1960s.

Model 5500

Period
Early 1960s

Movement
1520

Bracelet
19mm - rivetted
construction

During this period Rolex began to change the wording at the bottom of the dial from *Swiss* to *Swiss T<25*. On some watches *T<25* was printed in white on either side of the word *Swiss* and these dials also had a white line over-printed below the words *Super Precision.* This was a registration mark that ensured that *T<25* was correctly positioned when it was added to the dial.

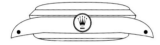

Model 1016

Period
Early 1960s

Movement
1570

Bracelet
20mm - rivetted
construction

The second version of the 1016 model had an all-white print dial. The word *Rolex* was written in a slab-serif typeface.

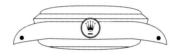

81

RLX

The serif typeface used to write *Rolex* on the dial of 1016 models from the mid 1960s.

Model 1016

Period
Mid 1960s

Movement
1560

Bracelet
20mm - rivetted construction

The Space-Dweller was a version of the Explorer, and was launched in 1963 to commemorate a visit to Japan by the astronauts from NASA's Mercury Space Program.

The production run for these watches was small, and they were not released into all Rolex's world markets.

Model 1016

Period
Mid 1960s

Movement
1570

Bracelet
20mm - rivetted construction

On the third version of the 1016 model, the word *Rolex* is written in a serif typeface on the dial.

The watch was now powered by the 1570 movement.

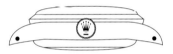

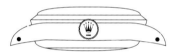

	1963
	Space-Dweller is launched

Model 1016

Period
Early 1970s

Movement
1570

Bracelet
20mm - folded
construction

In the early 1970s, the shape of
the Rolex crown logo changed.
Compare this example with the
Rolex crown logo printed on the
previously shown 1016 model.

Model 5500

Period
Early 1970s

Movement
1520

Bracelet
19mm - rivetted
construction

On the final version of the Air-
King, model 5500, the word
Precision was written on the dial.
Before both *Precision* and *Super
Precision* had been used.
  During this period the watch
was issued with either a rivetted
or folded Oyster bracelet.

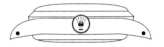

Explorer

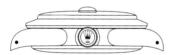

The distinctive bezel used on the 1655 model.

Model 1655

Period
Early 1970s

Movement
1575

Bracelet
20mm - rivetted construction

The Explorer II, model 1655, was launched in 1971.

On the first versions there was no luminous circle on the seconds hand.

Model 1655

Period
Mid 1970s

Movement
1575

Bracelet
20mm - rivetted construction

The second version of the 1655 model had a luminous circle on the seconds hand.

1971
1655 model launched

Model 1655

Period
Mid 1970s

Movement
1570

Bracelet
20mm - folded
construction

On the third version of the 1655
model the extra hour hand was
painted red. Previously it had
been painted orange.

This model was fitted with
identical movements marked
either 1570 or 1575.

Model 1655

Period
Late 1970s

Movement
1575

Bracelet
20mm - solid
construction

On the dial of the final version of
the 1655 the word *Rolex* is written
in a serif typeface. Previously it
was written in a slab-serif typeface.

The shape of the Rolex crown
used on the dial was different
to the one used on the previous
model.

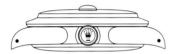

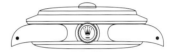

Explorer

The bezel used on
the 16550 model.

Model 1016

Period
Early 1980s

Movement
1570

Bracelet
20mm - solid
construction

During the 1980s the Mercedes
hands used on the 1016
changed. The new hands were
larger, and on the seconds hand
the luminous circle was moved
further down the shaft. The
larger size meant that the minute
and seconds hands overhung the
minute track.

Model 16550

Period
Mid 1980s

Movement
3085

Bracelet
20mm - solid
construction

In 1985 the 16550 model replaced
the 1655 model. It was fitted with a
synthetic-sapphire crystal and was
available with a white or black dial.

The white dial watches have
become very collectible - the paint
used was a pale ivory color, which
faded to a distinctive cream that
collectors have found especially
attractive. The first models have
white gold hands and white gold
settings around the dial's luminous
markings

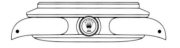

1985
16550 model launched

Model 1016

Period
Late 1980s

Movement
1570

Bracelet
20mm - solid
construction

The final version of the 1016
model had a flatter lens profile
than before.

Chapter 7

Milgauss

Created in 1954, the Milgauss was the first anti-magnetic wristwatch, designed for use in power stations or other conditions where high magnetic fields existed.

Milgauss

Introduced in 1954, the Milgauss was designed for use in high magnetic fields. It was primarily created for people working on scientific experiments or in power stations, who would need a watch that remained accurate in conditions of work where high magnetic fields existed. The watch's distinctive name was derived from the French word *mille*, meaning one thousand, and gauss, which is a measurement of magnetic induction.

Magnetic fields, measured in oersted, affect the balance workings of a watch. A normal watch will become erratic at 60 oersted and cease to function at around 100 oersted. The first Milgauss, model 6541, had anti-magnetic alloys in parts of its movement, which was encased by an iron shield, thereby enabling the watch to remain accurate in fields of over 1000 oersted. The movement used was a modified 1065 calibre, marked 1065M, and later, the 1066 movement was used, also marked as the 1066M.

At the time of its launch the Milgauss was considered a marvel of horological engineering. Not only did the watch function perfectly when subjected to a magnetic field of 1000 oersted (its guaranteed limit), but it could also remain accurate at 5000 oersted. In fact, even after being placed in such a powerful magnetic field, the residual magnetism in the watch's workings was insufficient to affect its accuracy. One might say that this is even a feat of over-engineering - rather like taking a Submariner depth-rated to 200 meters (660 feet), down to 1200 meters (3960 feet), with the knowledge that it would still be perfectly waterproof on its return.

The Milgauss went through a period of rapid change in the first few years of its existence. There is a common misconception that the Milgauss was launched with Mercedes hands and the distinctive lightning bolt seconds hand, but these were actually the last modifications made to the watch in the late 1950s.

The first version of the 6541 model had leaf-shaped hands, a Submariner bezel and what was probably a Submariner dial. The word *Milgauss* was written in red above the six o'clock position. A very small number of these watches was produced before being superseded by the second version, which now had a dedicated Milgauss dial, and displayed the word *Milgauss* in what subsequently

The Milgauss, model 6541, launched in 1954.

The third version of the Milgauss, model 6541, which had a unique bezel.

became the usual position: below twelve o'clock. There were also metal arrowhead markers at the three o'clock, six o'clock and nine o'clock positions, which had small luminous spots next to them. Like the Turn-O-Graph and Explorer, the early Milgauss models were available with a honeycomb dial.

The third version of the 6541 introduced a unique Milgauss bezel, which was marked from one to five, and had a red triangle at the top of the bezel, above the twelve o'clock position on the dial. The seconds hand also acquired a small red triangle, and the luminous spots beside the arrowhead markers disappeared. On the fourth version of the watch, the lightning bolt seconds hand was introduced, followed rapidly by the Submariner's Mercedes hands. After this last change, the 6541 now looked very similar to the Submariner, and perhaps because of this it did not sell in great numbers.

In the early 1960s a new model, the 1019, replaced the 6541. The 1019 model came with a solid metal bezel and an aluminium dial in either brushed or natural metal, or black. It was originally fitted with the 1080 movement, and

offered many different combinations of baton markers in either solid steel or in-filled with luminous or black paint. Batons made of solid metal or those in-filled with black paint had a small luminous spot beside them. The early versions of the 1019 model display quarter and half seconds, as well as seconds, on the dial. Versions that appeared from the mid-1970s onwards display half seconds and seconds. On the early versions the hands are made of an anti-magnetic alloy, and have an angled profile, whilst on the later versions, the hands are made of aluminium cut from a flat sheet.

It would seem that neither the 6541 nor the 1019 models were ever great sellers. The Milgauss had a small professional market, and its other buyers would probably have just liked the look of this large-case watch, and never required its unique anti-magnetic features. The Milgauss was finally withdrawn from the Rolex catalog in 1988, although for a short time after this it was still available as a special order piece.

The Milgauss, model 1019, which replaced the 6541 model in the early 1960s.

Milgauss

The first versions of the Milguass had a Submariner bezel.

Model 6541

Period
Mid 1950s

Movement
1065M

Bracelet
20mm - rivetted construction

The 6541 model was the first Milgauss, launched in 1954.

On the first versions, the word *Milgauss* was written in red above the six o'clock position. The watch was fitted with a Submariner bezel and had leaf shaped hands.

Very few examples of this version were produced before it was replaced by a second version.

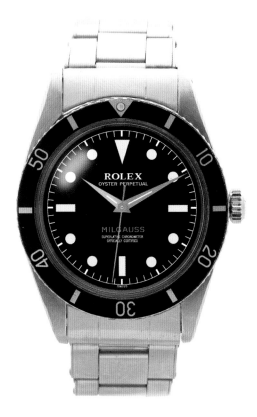

Model 6541

Period
Mid 1950s

Movement
1065M

Bracelet
20mm - rivetted construction

The second version had a dedicated Milgauss dial, and displayed the word *Milgauss* in what subsequently became the usual position: below twelve o'clock. The dial had metal arrowhead markers at the three, six and nine o'clock positions, which had small luminous spots next to them.

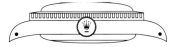

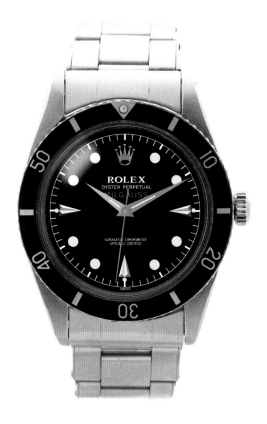

Key dates	**1954**
	6541 model launched

The third Milgauss launched had a dedicated Milgauss bezel which was marked from one to five, and had a red triangle at the top.

Model 6541

Period
Mid 1950s

Movement
1065M

Bracelet
20mm - rivetted construction

On the third version there was a unique Milgauss bezel and the seconds hand acquired a small red triangle at the tip. On the dial, there were no luminous spots beside the arrowhead markers.

Model 6541

Period
Mid 1950s

Movement
1066M

Bracelet
20mm - rivetted construction

On the fourth version of the watch, the lightning bolt seconds hand was introduced.

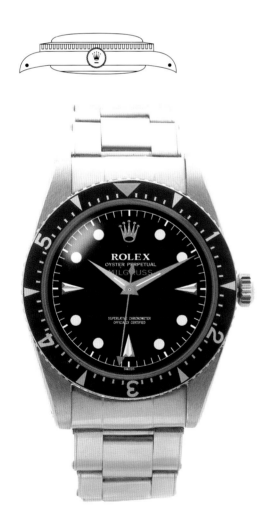

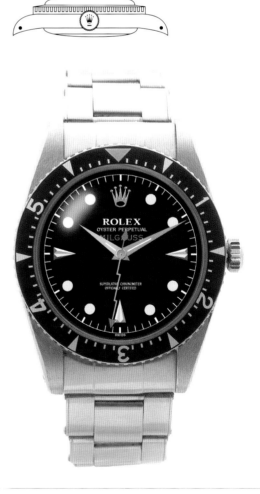

Milgauss

Model 6541

Period
Mid 1950s

Movement
1066M

Bracelet
20mm - rivetted
construction

This example has a textured dial,
described by Rolex as a
honeycomb dial, which was fitted
to many of the early examples of
this model.

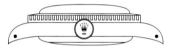

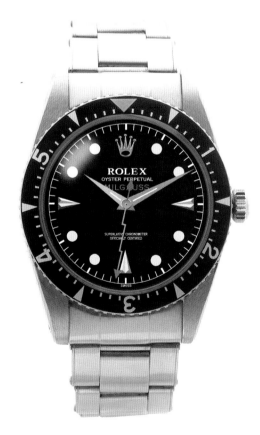

Model 6541

Period
Late 1950s

Movement
1066M

Bracelet
20mm - rivetted
construction

This example shows the 6541
model in its final form, fitted with
Mercedes hands. This watch was
replaced by the 1019 model.

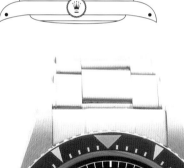

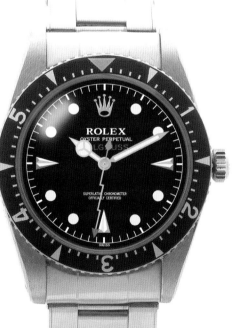

Model 1019

Period
Early 1960s

Movement
1080M

Bracelet
20mm - rivetted
construction

In the early 1960s, the 1019 Milgauss replaced the 6541 model. The watch came with an aluminium dial in either natural metal or black. The early versions of the dial display quarter and half seconds, as well as seconds.

On the first versions the hands were made of an anti-magnetic alloy, and had an angled profile.

This example features a dial which has metal baton markers in-filled with luminous paint.

Model 1019

Period
Early 1960s

Movement
1080M

Bracelet
20mm - rivetted
construction

On this example the metal batons are in-filled with black paint. Metal batons that were not in-filled with luminous paint had a luminous spot beside them.

Both this dial and the one previously illustrated were available until the watch ceased production.

c.1961

1019 model launched

Model 1019

Period
Late 1960s

Movement
1580M

Bracelet
20mm - rivetted
construction

On this dial both the hands and
baton markers are in-filled with
black paint, and there are semi-
circular luminous markings
beside them.

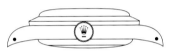

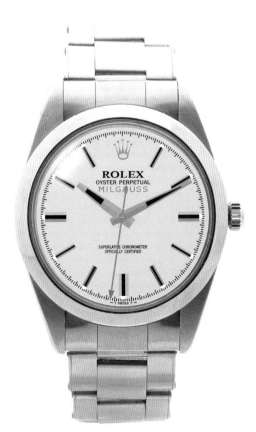

Model 1019

Period
Mid 1970s

Movement
1580M

Bracelet
20mm - folded
construction

From the mid-1970s, the 1019
dial only displayed half seconds
and seconds, and the hands
were made of aluminium cut from
a flat sheet.
   On this dial the baton markers
are in-filled with black paint, and
there are semi-circular luminous
markings beside them.

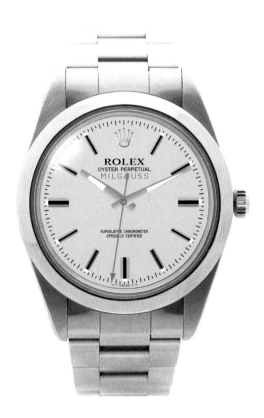

Model 1019

Period
Mid 1970s

Movement
1580M

Bracelet
20mm - solid
construction

On this dial the baton markers
are solid steel, and there are
semi-circular luminous markings
beside them.

Model 1019

Period
Mid 1970s

Movement
1580M

Bracelet
20mm - solid
construction

This example has an alternative
dial which has circular luminous
spots beside the metal batons. It
has solid steel baton markers
which are not in-filled with either
luminous or black paint.

Chapter 8

Turn-O-Graph

Created in 1953, the Turn-O-Graph was a tough, waterproof watch which could record elapsed time.

Turn-O-Graph

In 1953 Rolex launched the Turn-O-Graph, model 6202. It featured a rotating bezel to measure elapsed time, known by Rolex as a time-recording rim. On the bezel the minute divisions were shown as small circles, and every tenth minute was marked with a numeral. This first Turn-O-Graph had pencil-shaped hands, and the luminous circle was placed at the tip of the seconds hand. It was depth-rated to 50 meters (165 feet) and initially used the A296 movement. It was later fitted with the A260 movement.

On the first models there was no luminous bubble at the twelve o'clock position, and the circular minute divisions were in the center of the bezel. On later models, the minute marks were placed nearer to the bezel's inside edge.

In the mid-1950s, both the Turn-O-Graph's and the Submariner's hands were changed to Mercedes hands, and the luminous circle on the seconds hand was moved further down the shaft. In keeping with all sports models of this period, when the new seconds hand was introduced, it was painted white.

The first few dial designs are very distinct. *Turn-O-Graph* is written in extremely small lettering beneath the Rolex signature, and the words *Oyster* and *Perpetual* are separated by a large gap. It looks very much as if Rolex was still experimenting with where to place this information on the dial. A year after its launch, the name *Turn-O-Graph* appeared in a larger size, in the more familiar position above six o'clock. Like the early Explorers, the Turn-O-Graph was available with a textured dial, described by Rolex as a honeycomb dial, to soften its austere looks.

A deluxe version of the watch was also available with a solid gold bezel, and white or black honeycomb dial. Very few of these watches are ever seen now, and it is likely that the deluxe version was a special order piece, made only in small numbers.

Like the later Submariner, the Turn-O-Graph was marketed for its ability to accurately record elapsed time, thereby

The Turn-O-Graph, model 6202, launched in 1953.

obviating the need for an expensive chronograph. The fact that it was waterproof was also emphazised, because at that time Rolex had not yet produced a waterproof chronograph.

On the cover of the first Turn-O-Graph promotional brochure, the watch was shown on the wrist of a male airline passenger, thus presenting it as the ideal timepiece for the international business traveller. This market positioning was superseded two years later by the 1954 launch of the GMT-Master, which, although it was aimed specifically at airline crews, was purchased by crew and passengers alike.

The Turn-O-Graph continued in production until the early 1960s. Its eventual demise was due partly to the popularity of the GMT-Master, and partly to the continuing improvement in the Submariner's depth rating. Because of the similarity between these three models there was little reason to buy the less depth-rated Turn-O-Graph.

The Turn-O-Graph, along with the Milgauss, is the only sports model no longer in production, but they remain important parts of the history of Rolex's sports models. The Turn-O-Graph appears to be underrated by current collectors, but its small size and distinctive bezel make it a very attractive watch.

The final version of the Turn-O-Graph, model 6202, which had a luminous bubble at the twelve o'clock position.

Turn-O-Graph

On the first bezel there was no luminous bubble at the twelve o'clock position, and the circular minute divisions were in the center of the bezel.

On the second version of the bezel the minute marks were placed nearer to the bezel's inside edge.

Model 6202

Period
Early 1950s

Movement
A296

Bracelet
20mm - rivetted construction

The Turn-O-Graph, model 6202, was launched in 1953 and depth rated to 50 meters (165 feet).

On the first versions of the dial *Turn-O-Graph* is written in extremely small lettering beneath the Rolex signature.

This example has a textured dial, described by Rolex as a honeycomb dial, which was fitted on many early examples of this model.

Model 6202

Period
Early 1950s

Movement
A296

Bracelet
20mm - rivetted construction

The Turn-O-Graph used a certified chronometer movement and the dial on the second version was marked *Officially Certified Chronometer*.

Key dates	**1953**
	6202 model is launched

Model 6202

Period
Early 1950s

Movement
A296

Bracelet
20mm - rivetted
construction

Not all Turn-O-Graph dials were
marked as an *Officially Certified
Chronometer*.

A chronometer designation is
not featured on the third version
released.

Model 6202

Period
Mid 1950s

Movement
A296

Bracelet
20mm - rivetted
construction

In the late 1950s, Mercedes
hands were fitted to the Turn-O-
Graph.

In keeping with all sports
models of this period, the
seconds hand was painted white.
On the first versions the luminous
circle is larger than on
subsequent versions.

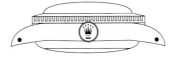

Turn-O-Graph

Model 6202

Period
Mid 1950s

Movement
A260

Bracelet
20mm - rivetted
construction

This example shows the second
version of the Mercedes hands,
which did not have the seconds
hand painted white.

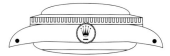

Model 6202

Period
Mid 1950s

Movement
A260

Bracelet
20mm - rivetted
construction

A few Turn-O-Graph versions
were produced with a depth
rating printed on the dial.

The final version of the Turn-O-Graph bezel had a luminous bubble at the twelve o'clock position.

Model 6202

Period
Late 1950s

Movement
A260

Bracelet
20mm - rivetted construction

This example is marked *Officially Certified Chronometer* on the dial.

Model 6202

Period
Late 1950s

Movement
A260

Bracelet
20mm - rivetted construction

On the final version of the Turn-O-Graph, the textured or honeycomb dial was no longer available. The watch continued in this form until the early 1960s when it was removed from the Rolex catalog.

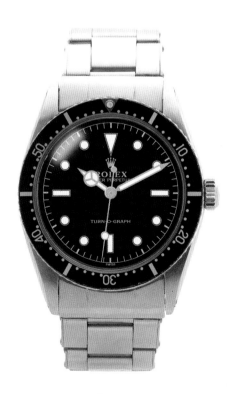

Chapter 9

Cosmograph

In 1960 Rolex launched the Cosmograph watch, designed to measure and record elapsed time.

Cosmograph

Rolex manufactured chronograph watches from the mid-1930s onwards. The first watch recognisable as the modern and collectible sports chronograph was the Cosmograph, model 6239, which was launched in 1960. (Rolex had previously made another watch named Cosmograph, which displayed the stages of the lunar cycle in a window in the dial, but despite having the same name, the 6239 did not have this lunar-phase feature.) The 6239 was powered by a 72B Valjoux movement, which Rolex had reworked and improved in its workshops. (Later the 722/1 movement replaced the 72B). The watch had a tachymetric timing scale on the bezel (the 6238 had had the tachymetric scale printed on the dial). On the first version of the 6239 the bezel was calibrated to 300 units per hour, but this was changed within a year to 200 units. It is not known why this change was made, but it certainly makes the bezel less cluttered and easier to read.

Very soon after the 6239 model was released, and was selling well in North America, Rolex often added the word

Daytona to the name (from Daytona Beach, Florida). Daytona Beach had been the home of car racing and testing for over a hundred years, and was known as the 'world center of speed'. It was a logical step for Rolex to promote this watch to the motor racing world, because it was a sophisticated chronograph, ideally suited both to reading lap times to within a fifth of a second, and for calculating average speeds. But for the collector it has contributed to confusion, because a single watch model might be marked *Cosmograph*, *Daytona*, or both, and in varying locations on the dial.

The 6239 was soon followed by the 6241 model, which was also fitted with the 72B movement. The only difference between the separate model numbers was the material used for the bezel. On the 6239 the solid bezel was of steel, with an engraved scale, and the 6241 had a steel bezel ring and a clear acrylic insert. This had the scale printed underneath, in white against a black background.

In 1965, the 6262 model, which had a

The Cosmograph, model 6239, launched in 1960.

The Cosmograph, model 6241. This example is fitted with a Paul Newman dial.

steel bezel, replaced the 6239. At the same time, the 6264 model, manufactured with an acrylic bezel insert, replaced the 6241. These new models were powered by the 727 movement. A little later, a new Cosmograph, the 6240 model, was launched. This watch introduced screw-down waterproof pushers, and therefore carried the word *Oyster* on the dial, denoting its waterproof status. The watch was depth-rated to 50 meters (165 feet) and was powered by the 727 movement. Neither the 6262 nor the 6264 models had waterproof pushers, and they were only produced for about four years before being replaced with the 6263 and 6265 models, which did have waterproof pushers

It was around this time, in the mid-to late 1960s, that NASA employees went incognito to a number of jewellers in Houston, Texas (where NASA was based), and paid cash for chronographs from five different watch manufacturers, including Rolex. The secrecy was important, because NASA did not want word to get out that the United States

space agency might be on the lookout for the watch which would be the first on the moon. All five watches were put through a series of gruelling tests, and unfortunately for Rolex and the Cosmograph, the only watch to pass all the tests was an Omega. (The chapter on the GMT-Master tells how all was not lost for Rolex - the company did get its watches into space, although never to the moon.)

In 1971 there were two Cosmograph models available: the 6263, which had an acrylic bezel insert, and the 6265 model, which had a metal bezel. On the later versions of the 6263 and 6265, the seconds hand was painted white on black-dial watches. Eighteen-carat gold versions of the 6263 and 6265 models were produced and tested to chronometer standard, and on the later versions these carry the words *Superlative Chronometer Officially Certified* on the dial. These two models were fitted with the 727 movement, and they remained in production until 1987, when the redesigned Cosmograph range replaced them.

The Cosmograph, model 6240. This watch introduced screw-down waterproof pushers.

Cosmograph

All the Cosmograph models were available with either standard black or silver dials. The silver dials had black recording dials, and on the black dials the recording dials were silver (with the exception of the gold 6263 and 6265 above). An extremely rare version of this standard dial exists, where both the overall dial and the recording dials are colored silver.

Rolex also made available a special dial known as the exotic dial. Today it is best known as the Paul Newman dial, having being so nicknamed by Italian collectors, after the American actor Paul Newman wore a watch with this dial for the cover photograph of an Italian magazine. The exotic dials were either black, on which the recording dials and the background for the seconds track were white, or they were cream with black recording dials and a black background for the seconds track. A different typeface was used for the numerals on the recording dials. The seconds track was printed in either white or red on a black background, or black or red on a white background.

The dials for both the standard and exotic models were signed *Daytona* or *Cosmograph* or with both names. The color and positioning of the name also varied - the words *Daytona* or *Cosmograph* appeared either in red or white print, depending on the countries in which the watch was sold. The name would be in a semicircle around the upper half of the lower recording dial, or below the twelve o'clock position, in a straight line. Today the exotic dial versions of the Cosmograph sell at a premium over the standard dials, and prices for these pieces have recently reached new highs. This is because the exotic dial is one of Rolex's most distinctive designs, fitted to a small and wearable watch.

The Cosmograph, model 6265, from the 1970s.

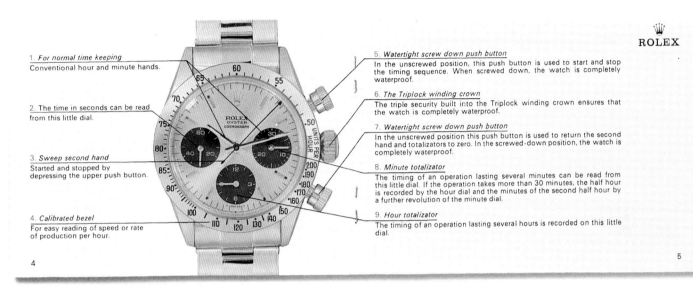

1. *For normal time keeping*
Conventional hour and minute hands.

2. The time in seconds can be read from this little dial.

3. *Sweep second hand*
Started and stopped by depressing the upper push button.

4. *Calibrated bezel*
For easy reading of speed or rate of production per hour.

5. *Watertight screw down push button*
In the unscrewed position, this push button is used to start and stop the timing sequence. When screwed down, the watch is completely waterproof.

6. *The Triplock winding crown*
The triple security built into the Triplock winding crown ensures that the watch is completely waterproof.

7. *Watertight screw down push button*
In the unscrewed position this push button is used to return the second hand and totalizators to zero. In the screwed-down position, the watch is completely waterproof.

8. *Minute totalizator*
The timing of an operation lasting several minutes can be read from this little dial. If the operation takes more than 30 minutes, the half hour is recorded by the hour dial and the minutes of the second half hour by a further revolution of the minute dial.

9. *Hour totalizator*
The timing of an operation lasting several hours is recorded on this little dial.

ROLEX

A page from a Cosmograph brochure showing the watches operational features.

The Cosmograph, model 6263, from the 1970s.

Cosmograph

The first version of the Cosmograph bezel was calibrated to 300 units per hour.

The second version of the Cosmograph bezel was calibrated to 200 units per hour.

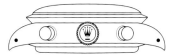

Model 6239

Period
Early 1960s

Movement
72B

Bracelet
19mm - rivetted construction

The 6239 model was launched in 1960. It was available with a standard dial, shown here, or with an exotic dial.

Model 6239

Period
Early 1960s

Movement
72B

Bracelet
19mm - rivetted construction

This example is fitted with the exotic dial, known to collectors as the Paul Newman dial, after the American actor Paul Newman wore this model on the cover of an Italian magazine.

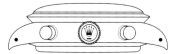

Key dates	**1960**
	6239 model launched

The Cosmograph bezel was also available with a plastic bezel insert.

Model 6241

Period
Early 1960s

Movement
72B

Bracelet
19mm - rivetted construction

The 6241 model was the second Cosmograph released. It had a steel bezel ring and a clear acrylic insert which had the scale printed underneath in white against a black background.

The watch was available with either the standard or exotic dial.

Model 6239

Period
Early 1960s

Movement
722/1

Bracelet
19mm - rivetted construction

There were numerous dial variations used in the Cosmograph range. On this example the word *Daytona* does not appear.

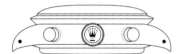

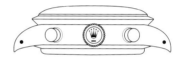

c.1961

6241 model launched

Cosmograph

Model 6239

Period
Early 1960s

Movement
722/1

Bracelet
19mm - rivetted
construction

The 6239 model was available
with an extremely rare version of
the standard dial, on which both
the overall dial and the recording
dials are colored silver.

Model 6241

Period
Early 1960s

Movement
72B

Bracelet
19mm - rivetted
construction

On several of the Cosmographs
fitted with a black dial the
seconds hand was painted white.

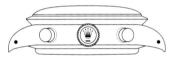

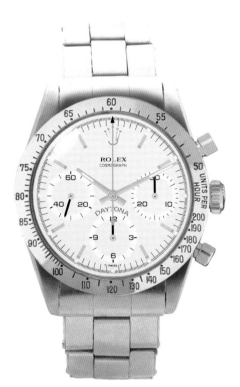

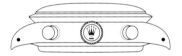

Model 6264

Period
Mid 1960s

Movement
727

Bracelet
19mm - rivetted
construction

In the mid 1960s the 6264 model
replaced the 6241 model.
　　The watch was available with
either the standard or exotic dial.

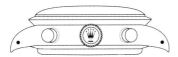

Model 6262

Period
Mid 1960s

Movement
727

Bracelet
19mm - rivetted
construction

In the mid 1960s the 6262 model
replaced the 6239 model.
　　The watch was available with
either the standard or exotic dial.

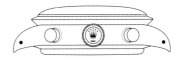

1965	**1965**
6264 model launched	6262 model launched

Cosmograph

Model 6240

Period
Mid 1960s

Movement
727

Bracelet
19mm - folded
construction

The 6240 model was launched in 1965. This watch introduced screw-down waterproof pushers, and carried the word *Oyster* on the dial. The watch was depth-rated to 50 meters (165 feet).

This model had a small winder and *O T Swiss T O*, at the bottom of the dial. It was available with either the standard or exotic dial.

Model 6265

Period
Early 1970s

Movement
727

Bracelet
19mm - folded
construction

In 1971 the 6265 model replaced the 6264 model. This model had a large winding crown and waterproof pushers.

The watch was available with either the standard or exotic dial.

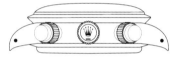

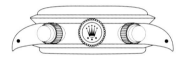

1965	1971
6240 model launched	6265 model launched

Model 6263

Period
Early 1970s

Movement
727

Bracelet
19mm - solid
construction

In 1971 the 6263 model replaced
the 6262 model.

The model was available with
either the standard or exotic dial,
and had *T Swiss T* at the bottom
of the dial.

On this example, the word
Cosmograph is written in a
semicircle around the upper half
of the lower recording dial.

Model 6263

Period
Mid 1970s

Movement
727

Bracelet
19mm - solid
construction

The 6263 model was available
with *Daytona* written in a
semicircle around the upper half
of the lower recording dial.

This model was available with
either the standard or exotic dial
and had *T Swiss T* at the bottom
of the dial.

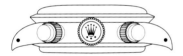

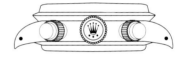

1971
6263 model launched

Chapter 10

Bracelets

The Oyster bracelet was patented in 1947, and was the only bracelet style to be factory-fitted to Rolex sports models until the mid-1960s, when the Jubilee bracelet was offered as an option on the GMT-Master. The Oyster bracelet was initially only available with straight ends but in 1954 the flush-fit endpiece was launched and was gradually fitted to all models. Over the years the Oyster bracelet has been constructed in three distinct styles: with either rivetted links, folded links or solid links.

The rivetted link design used hollow links attached together by round-ended rivets. Originally this bracelet was available with both fixed and expanding links for extra comfort, however problems arose when the expanding link bracelet was fitted to watches like the Submariner and Explorer, which were intended for strenuous use. If the bracelet was inadvertently snagged, its expanding links tended to allow the watch to be easily pulled from the wrist. The springs inside each link also became loose over time, so the expandable version of the rivetted link bracelet was gradually phased out in the mid-1960s. It is very unusual today to find an expanding link bracelet in good condition, as they seldom survive as long as the watch they originally accompanied.

There are two styles of rivetted link bracelets that were created for different country markets. In Europe the center link is wider than the outer links, and the rivets are round-ended, whilst in the United States the central link is almost the same width as the outer links, and the rivets are round-ended with a hole in the center.

The folded link construction is simply surgical steel folded into the shape of the link. The edges of the link are given a satin finish and left visible. In solid link construction, the links are machined from one piece of steel. In earlier versions of this bracelet the links are thinner than on the later models.

These three bracelet styles did not directly replace one another as they were introduced, but were manufactured simultaneously for a time. Rolex also favoured certain bracelet designs in different regional markets, so it is possible to find a watch from the mid-1970s correctly fitted with any one of the three styles, depending on the market in which it had been originally sold. Broadly speaking, the rivetted link bracelet was in production from the late 1940s until the early 1970s, the folded link bracelet from the early 1960s till the early 1970s, and the solid link bracelet from the early-1970s to the present day.

The first clasp used on these bracelets had the Rolex crown at the very end of the buckle, where it formed a lip with which to open the clasp. The later fliplock clasp was designed for extra security, and was introduced in the late 1960s. At first it only appeared on the Sea-Dweller, and then on the Submariner, but it was never factory-fitted on any other watch.

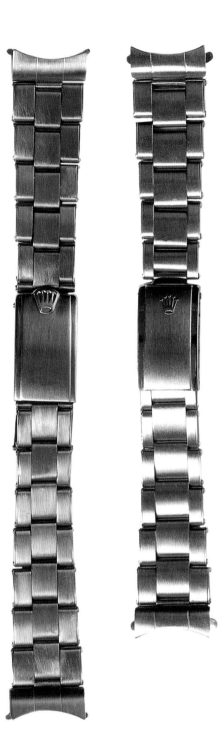

Below Left, Oyster bracelet with solid links.

Below center, Jubilee bracelet.

Below right, Oyster bracelet with a flip lock clasp. This bracelet was fitted to Submariners and Sea-Dwellers from the late 1960s.

Above Left, the first Oyster bracelet with straight ends. This bracelet was available with either fixed or expanding links.

Above center, United States Oyster bracelet which has a central link that is almost the same width as the outer links. This bracelet was available with either fixed or expanding links. This version is fitted with flush-fit endpieces.

Above right, European Oyster bracelet with a wider central link. This bracelet was available with either fixed or expanding links.

120

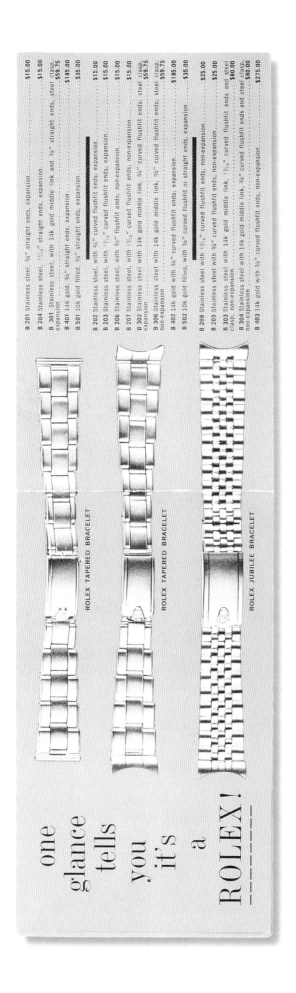

B 201 Stainless steel; ¾" straight ends; expansion............$15.00
B 204 Stainless steel; ¹¹/₁₆" straight ends; expansion............$15.00
B 301 Stainless steel, with 14k gold middle link and ¾" straight ends, steel clasp,
expansion............$59.75
B 401 14k gold, ¾" straight ends, expansion............$185.00
B 501 10k gold filled, ¾" straight ends, expansion............$35.00

B 202 Stainless steel, with ¾" curved flushfit ends; expansion............$15.00
B 203 Stainless steel, with 1³/₁₆" curved flushfit ends, expansion............$15.00
B 206 Stainless steel, with ¾" flushfit ends; non-expansion............$15.00
B 207 Stainless steel, with ¹⁵/₁₆" curved flushfit ends; non-expansion............$15.00
B 302 Stainless steel with 14k gold middle link, ¾" curved flushfit ends; steel clasp;
expansion............$59.75
B 306 Stainless steel with 14k gold middle link, ¾" curved flushfit ends; steel clasp;
non-expansion............$59.75
B 402 14k gold with ¾" curved flushfit ends; expansion............$185.00
B 502 10k gold filled, with ¾" curved flushfit or straight ends; expansion............$35.00

B 208 Stainless steel with 1³/₁₆" curved flushfit ends, non-expansion............$25.00
B 209 Stainless steel with 14k gold middle link, 1³/₁₆" curved flushfit ends and steel
clasp, non-expansion............$25.00
B 303 Stainless steel with 14k gold middle link, ¾" curved flushfit ends and steel
clasp, non-expansion............$80.00
B 304 Stainless steel with 14k gold middle link, ¾" curved flushfit ends and steel
clasp............$80.00
B 403 14k gold with ¾" curved flushfit ends, non-expansion............$275.00

ROLEX TAPERED BRACELET

ROLEX TAPERED BRACELET

ROLEX JUBILEE BRACELET

one glance tells you it's a ROLEX!

Rolex bracelet booklet
North American version, 1960.

On the left, are rivetted Oyster bracelets with either a straight end or a flush-fit endpiece. On the right is a Jubilee bracelet.
 The prices in the booklet are in US Dollars.

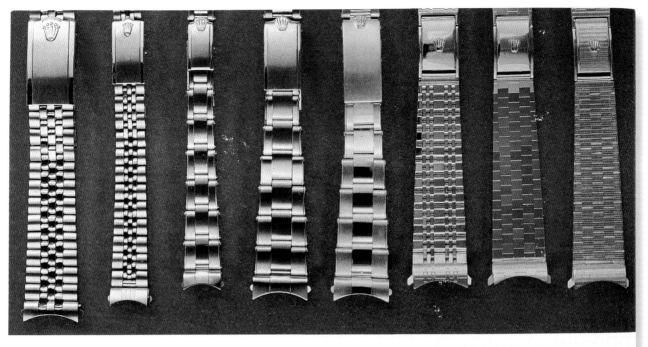

B200	B201	6634	6635	6635	2544	2557	2558
Stainless Steel	Stainless Steel	Stainless Steel	Stainless Steel	9ct. gold	9ct. gold	9ct. gold	9ct. gold

Distinctive bracelets designed specifically for Rolex Oyster watches. The "Flush-fit" ends eliminate the open spaces between the ends of the bracelet and the Oyster case. These bracelets are available separately.

32

Rolex booklet
European version, 1969.

Rivetted Oyster bracelets and Jubilee bracelets are shown.

Rolex box card

Expandable rivetted bracelets were susceptible to stretching. This card was issued to purchasers to warn them not to wind the watch whilst it was on their wrist because it increased the possibility of stretching.

Chapter 11

Boxes

The green and gold Rolex box, which usually comes with an illustrated outer cardboard box, is almost as distinctive as the watch within; and like the watches, the boxes have gone through constant development and change.

Boxes dating from the late 1940s to the early 1950s were long and narrow, allowing the watch to be displayed with the leather strap lying flat. These often had plain white outer cardboard boxes. In the early 1950s, when metal bracelets that did not open flat became more common, a cube-type box was adopted.

Boxes from the late 1950s, 1960s, and early 1970s had faceted tops, and came with dark green outer cardboard boxes. The inside of the box was of varnished wood, and contained a green padded felt card with two diagonal slots for the watch bracelet. By the late 1970s the faceted-top boxes were replaced by flat-topped boxes.

From the mid-1970s onwards the outer cardboard boxes changed from dark green to cream with a green stripe on one side, and by the late 1970s they changed again, now depicting seashells. A little later a pale green frost scene was in use. This design continued until the early 1990s. In 1976 it was the twenty-fifth anniversary of the Rolex Oyster, and all the boxes produced in that year had a commemorative medallion fitted to the interior.

By the 1990s a new box interior was in use, with the watch support changing from two slots to a soft pillow in cream suede that fitted into a rectangular recess. This box interior was used for most of the larger sports models, such as the Submariner and GMT-Master. Both styles were used for the Explorer 1016 model.

Original boxes can add value to a watch and should be stored with care in a safe place. Boxes from the 1970s are now being faked. The quality is not as good as that of an authentic box, so it pays to check them thoroughly. When buying a watch with a box, an effective way to check if the box is original to the watch is to look at the end of the outer cardboard box, which will have two paper stickers. One of these looks like a supermarket price tag, and displays the model number. The other is a circular sticker that should be of the same color as the watch dial color. The use of these stickers is a fairly late Rolex development, and only applies to watches sold from the 1970s onwards.

Paper sticker with the model number printed on. The model number is often written with a forward slash and a zero at the end, in this example, 1016 is written 1016/0. This was done in case Rolex modified the movement and needed to identify those versions with different movements, they would then replace the zero with another numeral.

Top left, box from the early
1950s which displayed the
watch lying flat. This box
could only be used for
watches fitted with leather
straps.

Top right, box from the late
1950s with a faceted top.
The inside of the box was of
varnished wood, and
contained a green padded
felt card with two diagonal
slots for the watch bracelet.

Bottom left, box from the
early 1970s with an angled
faceted top.

Bottom right, box from the
late 1970s with a flat top
and a metal badge.

Top left, flat topped box from the late1970s.

Top right, box from the late 1970s with a flat top which has a seam around the edge.

Bottom left, box from the early 1980s with a flat padded top.

Bottom right, box from the mid 1980s with a flat top and and no gold embossing.

Top left, outer cardboard
box from the 1950s.

Top right, outer cardboard
box from the late1960s and
early 1970s.

Bottom left, outer
cardboard box from the
late1980s and early 1990s.

Bottom right, outer
cardboard box from the
mid 1990s.

Chapter 12

Paperwork

In the art and antiques world collectible pieces are considered more valuable when they have a known provenance. For watch collectors, the equivalent of provenance is to have the original box and papers accompanying a watch. These could include the guarantee and timing certificate, and these can add up to forty percent to the value of the watch.

To find an original 6204 Submariner for which box and papers survive after forty-six years, is a very rare event. Less rare, but still exciting, is to find one of the first 1665 Sea-Dwellers from the early 1970s, with its original box and papers. The main market for the Sea-Dweller was the professional diver, who regarded it as a tool of his trade and therefore wouldn't have seen any reason to keep the packaging.

Papers can be easily checked, but the buyer should be aware that early paperwork often had only the model number printed on the guarantee. It is quite easy to add a 'unique' case number years afterwards.

Later paperwork uses a pinhole system, by which the case number is picked out in tiny holes. These are difficult (but not impossible) to fake. However the pinhole system was not in use worldwide.

It is always pleasant to come across a watch with supporting papers in addition to the timing certificate and guarantee. This might include the original receipt (and it can be depressing to see how much the first owner paid), the owner's brochure, subsequent service papers, a calendar, or other cards. The early cards that described the features of the watch and urged the buyer to throw away the card and use the box as a cigarette case are a true period touch – and since smoking was more widespread in the 1950s and 1960s, few of these cards and boxes survive

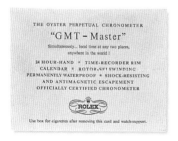

Box cards
English/North American versions, 1950s.

These cards described the features of the watch and urged the buyer to throw away the card and use the box as a cigarette case. Since smoking was more widespread in the 1950s and 1960s, few of these cards survive.

Left, a card for the GMT-Master, model 6542, which can be identified by the use of the designation *Officially Certified Chronometer*.

Center, card for the GMT-Master, model 1675. The designation *Officially Certified Superlative Chronometer* is a later version not used on the 6542 model.

Right, this is for the rarely seen Submariner 6200 model.

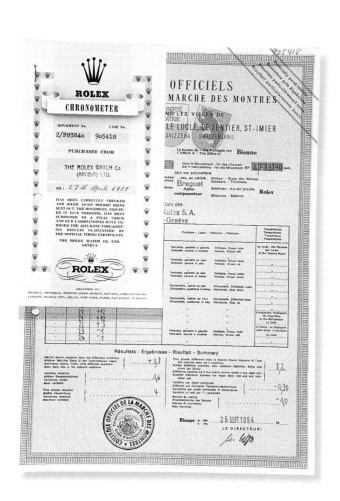

Chronometer certificate
1950s.

Far left, up until the 1960s the chronometer certificate was issued by the official testing center and had the individual test results entered by hand on the front.

Left, throughout the 1950s and 60s the number of watches submitted for testing grew and the large certificates were replaced by a certificate issued by Rolex.

Chronometer certificate
1970s.

This style of certificate was issued from the early 1970s onwards.
This example shows the pinhole system used to write the case number.

Chapter 13

Movements

North American Submariner movements

During the 1960s, American import tax was higher on watch movements fitted with over 20 jewels. Rolex created a version of the 1520 movement with seventeen jewels, rather than the normal twenty-six.

Submariner	Movement	Jewels	Period
Deep Sea Special/Piccard	1000/419343	21	Early 1950s
6205	A260	18	Early 1950s
6200	A296	18	Early 1950s
6204	A260	18	Early 1950s
6536	1030	25	Mid 1950s
6536/1	Chronometer version 1030	25	Mid 1950s
6538A	1030	25	Mid 1950s
6538	1030	25	Mid 1950s
5510	1530	26	Late 1950s
5508	1530	26	Late 1950s
5512	1570	26	Late 1950s
5512	1560	26	Early 1960s
5513	1530	26	Early 1960s
5513	1520	26	Mid 1960s
5513	North American version 1520	17	Mid 1960s
5517	1520	26	Late 1960s
1680	1575	26	Late 1960s
16800	3085		Mid 1980s

Sea-Dweller	Movement	Jewels	Period
5513 Comex	1520	26	Late 1960s
5514 Comex	1520	26	Late 1960s
1665	1575	26	Late 1970s
16660 Sea-Dweller	3035	27	Late 1970s
16660 Comex Sea-Dweller	3035	27	Late 1970s

GMT-Master	Movement	Jewels	Period
6542	1065	25	Mid 1950s
6542	1036	25	Late 1950s
6542	1066	25	Late 1950s
1675	1565	25	Early 1960s
1675	1575	26	Mid 1960s
16750	3075	27	Early 1980s

Explorer	Movement	Jewels	Period
6098	A296	18	Early 1950s
6298	A296	18	Early 1950s
6150	A296	18	Early 1950s
6350	A296	18	Early 1950s
6610	1030	25	Mid 1950s
1016	1530	26	Late 1950s
1016	1560	26	Late 1950s
1016 Space-Dweller	1560	26	Early 1960s
1016	1570	26	Mid 1960s
5500	1530	26	Late 1950s
5500	1520	26	Early 1960s
5504	1530	26	Late 1950s
1655 Explorer II	(marked 1570) 1575	26	Early 1970s
1655 Explorer II	(marked 1575) 1575	26	Early 1970s
16550 Explorer II with cream dial	3085	27	Mid 1980s

129

Milgauss	Movement	Jewels	Period
6541	1065M	25	Mid 1950s
6541	1066M	25	Late 1950s
1019	1080M	25	Early 1960s
1019	1580M	26	Mid 1960s

Turn-O-Graph	Movement	Jewels	Period
6202	A296	18	Early 1950s
6202	A260	18	Late 1950s

Cosmograph	Movement	Jewels	Period
6239	72B	17	Early 1960s
6239	722/1	17	Early 1960s
6241	72B	17	Early 1960s
6241	722/1	17	Early 1960s
6240	72B	17	Mid 1960s
6240	722/1	17	Mid 1960s
6262	727	17	Mid 1960s
6264	727	17	Mid 1960s
6263	727	17	Early 1970s
6265	727	17	Early 1970s

Chapter 14

Production dates

In order to identify (and to some degree, date) the ever-increasing numbers of watches being manufactured, Rolex's watch numbering system has become more complex over the years. The position of the numbers on the watchcase has also changed.

From the late 1940s Rolex used a circular case design. The model number was engraved between the lugs above the twelve o'clock position, with the individual case number between the lugs below the six o'clock position. By the early 1950s, Rolex had begun to add a date mark inside the watchcase. This showed the quarter in Roman numerals, followed by the last two digits of the year in Arabic numerals. For example, IV59 would date the watch to the fourth quarter of 1959. It can be appreciated that this system is problematic for collectors, because the vital date information can only be determined by opening the watchcase. In the mid-1950s, when production reached one million, Rolex did not continue to number watches in a chronological sequence, but instead started again at 100,000. In the late 1950s Rolex production increased, and when the second million mark was passed in the mid-1960s, the numbering system now carried on up to ten million.

When the ten millionth watch was made in the late 1980s, a new numbering system was introduced, which used a prefix letter followed by six digits (i.e. up to 100,000). The prefix letters used were not in alphabetical order, but instead started with the letter R

followed by L, E and X (the letter O from Rolex was not used as this would have been easily confused with the zero in the case number). Once the letter X had been introduced, Rolex began using the letters N, C, S, W, T, U. (Today Rolex's numbering system has changed again from those of the period covered by this book.)

The following are production dates of specific case numbers, which may help to date watches.

1950	673600
1955	135200
1960	140200
1965	1792000
1970	2831000
1975	4153100
1980	6203000
1985	8814000
1990	E000000
1995	T000000

The model number is engraved between the lugs above the twelve o'clock position.

The case number is engraved between the lugs below the six o'clock position.

During the 1950s Rolex began to engrave the inside of the case back with the model number and a date code which showed in Roman numerals the quarter of each year, followed by the last two digits of the year in Arabic numerals. This example shows IV65 which dates the watch to the fourth quarter of 1965.

On the case back engraving it is common to find a model number which does not match the watch case number. As many Rolex watches had identical backs, Rolex would often use case backs they had in stock rather than wait to have them engraved with the appropriate number, also, as model numbers changed old case backs with the previous model reference number on them were used up on new models, as it made no financial sense to destroy perfectly good case backs. Submariner, model 5512s were often fitted with 5513 model case backs, sometimes a 5512 case back has the model number crossed out and 5513 engraved below. This is also often seen on the case backs for early Explorers and Daytonas.

Chapter 15

Collecting

Over the last twenty years or so there has been significant and rapid growth in collecting wristwatches. With large well-established manufacturers like Rolex, which has been making watches for nearly a hundred years, there has been an accompanying growth of interest in the history and development of the company's watches. Today Rolex is recognised and esteemed by collectors for its innovation and important achievements in the horological field.

Creating a collection can be a tricky process. It may seem obvious, but the most important first step is to decide what to collect. The excitable collector will end up with all sorts of Rolexes, which make up a collection only insofar as they are a quantity of watches. If the goal is to develop a considered collection, it is important to choose a specific area on which to concentrate. This could be a certain period, such as the 1950s; or particular models, such as Submariners. What to collect will of course depend on the collector's budget, as well as interest.

Once a focus has been chosen, the level of condition at which to buy should be considered. If a large collection is the goal but budget is limited, then it may only be possible to purchase watches that have no paperwork, because watches that come with original paperwork can sell for thirty to forty

percent more than those with none. Original bezels add value, and original dials, even not in perfect condition, are worth more than refinished dials. Over-polished cases, where the metal has been worn away, detract from a piece's value.

No discourse on collecting Rolex watches would be complete without mention of the dreaded 'F' word - fakes. It was inevitable that the increase in serious collecting would be followed by an increase in faking. This began almost harmlessly, with poor quality fakes being sold on the streets of Bangkok and Hong Kong that bore little or no relationship to actual Rolex models. These were targeted towards the easy-going tourist in search of cheap and amusing souvenirs. But in the 1980s, the matter became serious when Rolex Prince model copies began to appear. These copies were intended to be good enough to fool both collector and dealer – which in some cases they did. This began the trend to forge collectible watches, thus insuring bigger profits for the counterfeiters.

Typically the most frequently faked watches were gold models, but with the growth in the value of sports models, even steel watches were faked. Before long, watch parts were being counterfeited as well. The Rolex Prince movement has a high value by itself, so

On the left, the first chronometer tag and, on the right, the second version. It is always fun to find watches which come with the original chronometer tags.

Gruen movements, which are similar, were altered by engraving the bridge with the Rolex signature.

Today the counterfeiters are producing better fakes of more and more models, but the collector need not be completely put off - there are some simple precautions which can minimise the risk of buying a fake.

• Identify and get to know reputable, knowledgeable and experienced dealers. Honest dealers usually do not mind recommending another dealer, so the collector can ask around and get to know - and stick to - the good ones.
• When you have a dealer you trust, don't hesitate to ask if the dial is refinished, or if the hands are original. A good dealer should know the answer and tell you.
• Never be rushed into a sale. A classic ploy is to pressure the buyer by saying the watch might be gone tomorrow. If the seller wants the money that quickly, it might be best to walk away.
• Fakes always seem to be sold with a tale of how it came to be in the seller's possession, so be wary of charming or convoluted stories intended to dispel doubt.
• Many fakes are tempting because they are offered for just a little less than the expected price. It is worth considering whether a reputable and knowledgeable dealer would really offer the watch in question at the quoted price.
• The growing popularity of internet selling offers the watch collector the benefit of being able to see many examples and gauge current prices. However, buying online is never the same as buying face-to-face, and the anonymity of the internet can offer an ideal home to both faker and con-man. Viewing a watch onscreen cannot reveal its true condition, nor does it substitute for a hands-on examination.

While there may be pitfalls of buying online, the internet can be a very beneficial research tool. A good net search will quickly reveal which dealers are selling a lot of watches online, and which of those have a good reputation. There are a number of specialist watch-related websites, and these may have areas where dealers and collectors can exchange information. There may also be information about fraudulent sales or untrustworthy dealers. Using these sites may help build up a good body of knowledge which might be later used to buy online with more confidence.

Here are examples of some current fakes:

• Exotic ('Paul Newman') dials are being copied and fitted to real Cosmographs. On genuine exotic dials, the recessed

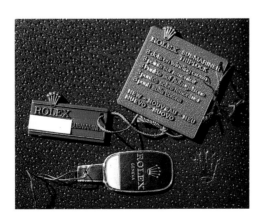

Original model tags from the 1960s and, on the right, the plastic tag which was issued with Submariner watches fitted with the very first trip lock winders.

recording dials have tightly-spaced concentric rings on their surface. On some of the fakes, these rings are widely spaced, and occasionally are absent altogether. The red paint used on some fakes is often warmer in tone than the original color.

• Counterfeiters overprint the word Submariner in red on the dial of the 1680 Submariner, which increases its value. Examining the red printing with a magnifying glass can help detect this. On fakes, red paint has been applied over the white paint and this can often be seen underneath where the red printing does not fully cover the white.

• Bezels for the first Submariner, Turn-O-Graph and Milgauss models are being faked. The typeface used is often incorrect, and the printing not as sharp as on the originals. A simple guide applies here. The bezel is a part of the watch that often gets scratched or worn - even just wearing the watch under a shirtsleeve can wear the bezel printing away after many years. So if the bezel on a watch over twenty-five years old is perfect, examine the bezel carefully against a genuine example.

• Service-use Submariners and Comex Submariners and Sea-Dwellers are being created from ordinary Submariners. On service-use Submariners, faked minute hands (or similar hands from other watch makes) are fitted. These hands tend to

be longer than the originals, and overhang the seconds track on the dial. With Comex Submariners and Sea-Dwellers, examine a few genuine examples to become familiar with the correct logo.

• Explorer dials are being fitted to Datejust models, which have the same size case as Explorers. If the watch is a genuine Explorer, its model number should be 1016.

• Explorer IIs are being created from GMT-Masters, which also have the same size case as Explorer IIs. The model number for a genuine Explorer II is 1655. When examining the watch, consider whether the number looks like it might have been filed down, and 1655 put in its place.

However collectors should not be disheartened. There are many more honest than untrustworthy or ill-informed dealers, so buying from reputable professionals and well-known auction houses should lessen any risks. When the collector has bought a watch (or better still, and if the dealer is willing, before the purchase has been made), he or she can take the watch to the nearest Rolex service center to have its authenticity checked. The service center will provide the collector with written details of the watch which also note the collector's name and address. Information can also be requested over

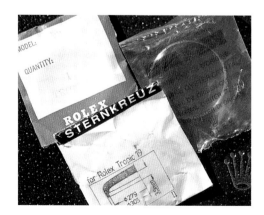

Original lenses from the 1960s and 1970s. More and more collectors want watches which are correct in every detail. Dealers are now selling old watch lenses to enable the collector to give a watch that important finishing touch.

134

the phone by giving the numbers of the watch and asking for a written confirmation of the details to be faxed.

Always get a receipt or statement of purchase that clearly shows the seller's name and the serial and model numbers of the watch. This need not be an invoice.

Because there are no official Rolex publications, it is worth noting that like collectors, dealers have built their knowledge through their own research. There are therefore no true experts, but there are certainly many opinions from which the collector can and should benefit before making up his or her own mind about a watch or its history. Although on occasion, misinformation or professional rivalry could mean that a genuine piece might be identified as a fake, on the whole the collector will find that listening and discussing, as well as researching, are invaluable tools to help build a significant collection.

Once you have built a collection remember to insure it and keep up to date on the value of your watches, so that you can adjust your insurance accordingly. A good way to keep track of the current market price is to register to receive catalogues from auction houses.

Remember to store your collection securely, a safe deposit box is probably the cheapest way of securing your collection, but it is often hard to find one that is near to your home. Having a safe

at home is a more costly option but allows you to be close to your collection, and you could use the extra space in the safe for storing other valuble items such as jewellery or important documents.

Finally, it would be inadvisable to collect Rolex watches to simply gain capital - there are less risky opportunities for financial growth. Any collection is best acquired through genuine interest and careful consideration, and even a small collection, if tightly focused, can command a greater value than the sum of its individual pieces. A collection should be treated as a portfolio of shares - the value of some pieces may fall, while others may rise. And unlike shares, a collection of Rolex watches is visually attractive and more directly enjoyable!

135

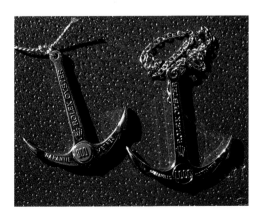

Original anchors from the 1960s and 1970s. These small metal anchors have the relevant depth rating on them. Having the correct anchor for a vintage Submariner watch is an added bonus.

Photo H. Broussard, Cannes

Style No 5512

A diver's dream come true!

The Rolex patented SPECIAL OYSTER case, dustproof and waterproof is specifically designed to be PRESSURE-PROOF to a depth of 660 feet under water.

Revolving "Time Recorder" on outer bezel; milled edges for easy manipulation. Calibrated figures clearly marked on black background. Rolex patented self-winding, 25 jewels, Super-Shock resisting. Radiation-safe. Winding crown protected by steel shoulders.

Another version of this model, waterproof to a depth of 330 feet and without the protective crown shoulders, is available under ref. 5508.

ROLEX official timepiece "Pan American Airways"

THE ROLEX WATCH CO. LTD. - GENEVA, SWITZERLAND
(Founder, H. Wilsdorf)

ADEN, BOMBAY, BRUSSELS, BUENOS AIRES, COLOGNE, DUBLIN, HAVANA, JOHANNESBURG, LONDON, MEXICO CITY, MILAN, NEW-YORK, PARIS, SAO PAULO, SINGAPORE, TOKYO, TORONTO.

Chapter 16

Advertisements & Literature

The Illustrated London News
Magazine advertisement, 1954.

One of the first advertisements to show the new Submariner watch. The watch featured could be the 6204 or 6205 model. The watch was priced at £66/$99.

138

I'm glad I wore my waterproof watch today

When it's raining cats and dogs, you'll be glad you have a waterproof Rolex Oyster on your wrist. On more gentle occasions at business and at home you'll feel *correct* with a slim, elegant Rolex dress watch on your wrist. If you are unable to call to see the largest selection of Rolex watches, may we send you our catalogue and details of our personal service, which includes free insurance and a two year guarantee against *everything*, including accidental damage.

ROLEX SUBMARINER

Stainless steel waterproof case, 25-jewel "rotor" self-winding movement. Extra luminous dial and hands. The revolving bezel acts as a constant reminder of a set time. £66. 10s. 0d. *Other Oyster watches from £25.*

Charles Packer
76 REGENT STREET
LONDON, W.1

ESTABLISHED 1787   REGent 1070

from the
TOP OF
THE WORLD

to the
BOTTOM
OF THE SEA

ROLEX *proves dependable!*

In the same year when the roof of the world was

at last conquered by the Rolex equipped Everest expedition,

the Rolex Company produced a special "Oyster" model watch which, affixed outside

the Bathyscaphe "Trieste," submerged to 10,350 feet . . .

and after surfacing was in perfect condition . . . keeping perfect time.

With the new-found sports craze . . . "skin-diving" . . . and as a result of deep-sea testing, Rolex is now able to add to the equipment of skin-divers with a unique, new water-proof, pressure-proof watch . . . the "Submariner." This instrument has a special rotating bezel, enabling the user to determine elapsed time at a glance: for instance, it keeps a diver alert to how much time he has before his supply of air will run out. The "Submariner" has almost all the advantages of a stop watch with none of the complications. Unconditionally guaranteed against any water pressure . . . here is another truly incomparable Rolex masterpiece.

BE SURE TO WRITE FOR THE SPECIAL FREE BOOKLET ON THE ROLEX "SUBMARINER"

ROLEX
A Landmark in the History of Time Measurement

THE AMERICAN ROLEX WATCH CORPORATION · 580 FIFTH AVENUE · NEW YORK 19, N. Y.

Mention the National Geographic—It identifies you

National Geographic
Magazine advertisement, 1953.

Advertisement for the Submariner which describes the functions of the new watch.
   The watch shown is an artist's impression of the Submariner, created by using a drawing of one of the Deep Sea Special watches. The artist has added a drawing of a Turn-O-Graph bezel to create an odd-looking watch which Rolex certainly never manufactured.

The London Times, color supplement
Magazine advertisement, 1953.

Rolex was justifiably proud of the use of its watches on the successful Everest expedition of 1953. The drawing at the bottom of the page shows the type of watch issued to the expedition and confirms that they were standard Oyster Perpetuals, with white dials which did not display *Explorer* on the dial.

Everest Leader's Tribute to Rolex

ON MAY 29TH, 1953, the British Everest Expedition, led by Brigadier Sir John Hunt, finally reached the summit of Mount Everest. Rolex Oyster Perpetual watches were supplied to the expedition. Sir John pays this tribute to Rolex.

"The Rolex Oyster Perpetual watches, with which members of the British team were equipped, again proved their dependability on Everest. We were delighted that they kept such accurate time. This ensured that synchronisation of time between the members of the team was maintained throughout.

"And the Oyster case lived up to its reputation, gained on many previous expeditions, for protecting the movement. Our Rolex Oysters were completely waterproof, unharmed by immersion in snow, and withstood the extreme change of temperature from the warm humidity of the foothills to the great cold at the high camps.

"Last, but not least, the Perpetual self-winding mechanism relieved the team from the trouble of winding their watches. At heights of over twenty-five thousand feet this is really necessary, because the mind slows up and such details as winding watches can be forgotten. There was no need either to slip off warm gloves to attend to this detail.

"As I have emphasized before, this expedition was built on the experience and achievement of others. Rolex Oyster watches have accompanied many previous pioneering expeditions. They performed splendidly, and we have indeed come to look upon Rolex Oysters as an important part of high climbing equipment."

John Hunt.
Leader

15th June, 1953
Khatmandu

THE ROLEX EXPLORER—a new watch built specially for scientists and explorers to withstand every conceivable hazard. The famous Oyster waterproof case has been strengthened to stand up to tremendous pressures. The Explorer functions perfectly to a depth of 300 feet under water and to a height of 12 miles. It is wound automatically by the unique Rolex Perpetual self-winding "rotor" which, by keeping an even tension on the mainspring, ensures the utmost accuracy. The Explorer is anti-magnetic. It has highly luminous dial-figures on a jet-black dial. It costs £49.19.6d., including the steel bracelet.

As from now, all Rolex Oyster Perpetuals are graced by the new, slimmer Oyster case. This unique invention infallibly protects the movement against water, dust and damp. It is guaranteed to withstand temperatures from 10° F to 180° F (−25° C to +82° C) and to resist pressure to a depth of 150 feet (50 metres) under water.

The smooth-running, silent, self-winding "rotor" keeps the Rolex Oyster Perpetual fully wound automatically, if the watch is worn for as little as 6-8 hours a day. This constantly even tension on the mainspring makes for still greater accuracy.

ROLEX

A landmark in the history of Time measurement

THE ROLEX WATCH COMPANY LIMITED (*H. WILSDORF, GOVERNING DIRECTOR*),
1 GREEN STREET, MAYFAIR, LONDON, W.1

The Illustrated London News
Magazine advertisement, 1954.

At the bottom left of the advertisement is an Explorer with a black dial. It is almost certainly a 6150 model and features an early version of Mercedes hands.

The watch is described as a new model created for scientists and explorers. It also states that all Oyster Perpetuals now had a thinner case design - the thicker bubble-back case was no longer in use.

The watch was priced at £49/ $73.50.

141

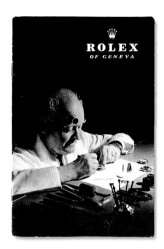

Hands   **alone ...**

Here is horological science at its purest—the surety of touch and the clarity of sight. *Top left*, assembly of parts of a watch; some parts are so small that they literally cannot be seen in this photograph—

Top right, oiling the trains by hand, a job calling for great delicacy— *Bottom left*, a young girl checking a hairspring— *Bottom right*, a mainspring being wound preparatory to being introduced into its housing.

Guide to the Rolex factory
English/North American
version, 1959.

The delicate nature of the work undertaken at the Rolex factory was tightly controlled, and no visitors were permitted. Instead, this brochure was sent out to people who had requested a tour.

Above, movements in various stages of assembly are shown.

Below, the highly accurate clock used to set and monitor the timekeeping for non-chronometer watches and the exterior of the Rolex factory.

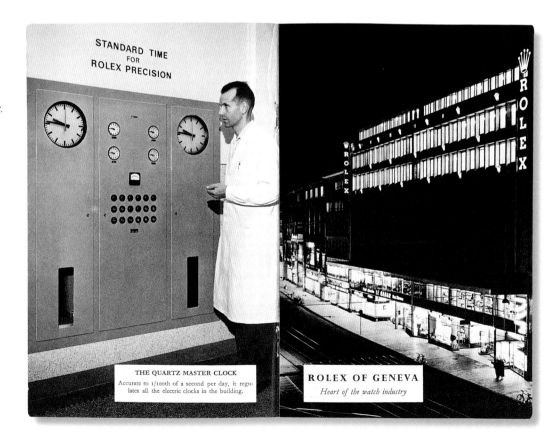

STANDARD TIME FOR ROLEX PRECISION

THE QUARTZ MASTER CLOCK
Accurate to 1/100th of a second per day, it regulates all the electric clocks in the building.

ROLEX OF GENEVA
Heart of the watch industry

Above: Pressure gauges, waterproof gauges, fathometers and special chucks—all used for the testing of the OYSTER case and all invented and patented by Rolex. The operator is using a special machine to unscrew the back of an OYSTER case.

Below: A glimpse of just a few of the special tools used for the Rolex OYSTER case—the case that is unconditionally guaranteed 100% waterproof to a depth of 165 feet below water.

A special gauge for testing the effects on a watch of pressures up to 600 atmospheres—or four miles under the sea! On the right is the famous Rolex OYSTER which was fitted to the *outside* of Professor Piccard's bathyscaph which dived to a record depth of 10,335 feet, where pressure is almost two tons per sq. inch. The watch worked perfectly! This special model was produced at considerable expense merely to demonstrate to ourselves whether we were able to cope with such a problem.

Above, the factory tool area is shown with its vast selection of tools, used to open various Rolex case backs. Also shown is a watch is described as one of the Deep Sea Special watches which were created in 1953, however, from the design of the winding crown and hands it looks more like a Piccard model which was created in 1958. A Rolex technician is shown operating the specially created pressure testing device used to verify these watch's waterproofing to an incredible 600 atmospheres of pressure.

Below, is a machine which removed tiny amounts of acrylic from Rolex lenses, to ensure an accurate and waterproof fit.

PRECISION . . . Rolex precision calls for gauges, micrometers and measuring instruments of *exceptional* accuracy, for chronometric exactitude depends on tolerances a hundred times thinner than a human hair.

Right: Piercing the hole in the watch case which will receive the stem of the winding crown—accuracy here must be to 1/100th of a millimetre.

Below: Tools and gauges being manufactured exclusively for Rolex. It is the policy of insisting on the special manufacture of precision instruments that has helped place Rolex in the forefront of those who measure Time.

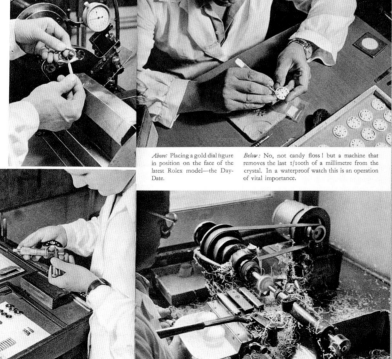

Above: Placing a gold dial figure in position on the face of the latest Rolex model—the Day-Date.

Below: No, not candy floss! but a machine that removes the last 1/100th of a millimetre from the crystal. In a waterproof watch this is an operation of vital importance.

TO SERVE YOU BETTER

144

FACTS YOU SHOULD KNOW ABOUT YOUR ROLEX WATCH

THIS IS A " JOB TICKET ". It is to us an important part of your watch, for after careful inspection our shop foreman indicates *on the job ticket itself* each of the important parts of your watch which require attention.

Every trade has its own vocabulary. We in the watch business sometimes overlook the fact that you, our customer, may not be familiar with our particular jargon. We feel that by learning the meaning of our most oft-used terms, you will at the same time learn something more about your watch.

By better understanding the need for fine care of your watch, you will perhaps better appreciate the necessity for the valuable service we can render you.

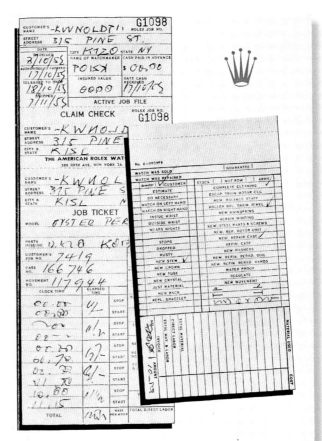

**THE AMERICAN ROLEX WATCH
CORPORATION**

**580 Fifth Avenue
NEW YORK 36, N.Y.**

SAFETY CHECK LIST

When a sport places its participant out of his natural element, Safety First becomes doubly important. In the interests of safety and as a convenience for SCUBA* diving enthusiasts, Rolex offers this booklet of hand signals and the important Navy Standard Decompression Table.

Newcomers to SCUBA diving will wisely limit their dives to a maximum of 33 feet at first, since at this depth decompression is unnecessary. (However, never ascend faster than 25 feet per minute.)

For the experienced SCUBA diver, thorough knowledge of the decompression table lessens the risk of getting "the bends" following deeper dives. The Rolex Submariner watch with its elapsed time indicator is a reliable and necessary companion on such dives!

* Self Contained Underwater Breathing Apparatus. You eager beginners who may want to go off the deep end with SCUBA — don't — until you've mastered the basics with a diving mask and instructor!

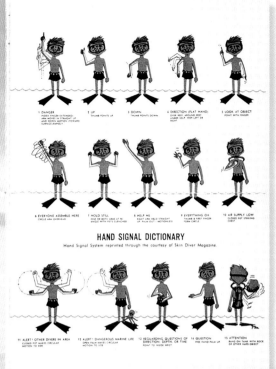

HAND SIGNAL DICTIONARY
Hand Signal System reprinted through the courtesy of Skin Diver Magazine.

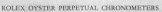

ROLEX OYSTER PERPETUAL CHRONOMETERS
Where dependability, ruggedness and absolute accuracy are imperative you will always find Rolex Oyster Perpetual Chronometers. Hermetically and permanently sealed against the elements... waterproof*, dustproof, sandproof and perspiration-proof... the Rolex Oyster case gives the Officially Certified Chronometer movement protection to assure complete accuracy under all conditions.
* When case, crown and crystal are intact.

THE ROLEX SUBMARINER
... specifically designed for deep sea diving... pressure-proof to 660 feet underwater. The rotating bezel tells elapsed submersion time at a glance. Stainless steel Oyster case with 25 jewel self-winding Perpetual chronometer movement... black dial with prominent luminous dots and markers. Winding button protected by steel shoulders.

ROLEX — THE WATCH WORN BY FAMOUS UNDERWATER EXPLORERS THE WORLD OVER!

© 1961 Montres Rolex S.A.   PRINTED IN SWITZERLAND   11 M · 254 · 504

Skin Diver Handbook
North American version, 1961.

Left, the back cover shows a 5512 Submariner with the early pointed profile crown guards. These were also fitted on the first GMT-Master, model 1675.

The brochure opened out to reveal a set of standard American Naval decompression tables (not shown).

GMT-Master owner's booklet
English/North American
version, 1960.

The GMT-Master, model 6542, is featured. This model did not have the protective crown guards that were fitted to the 1675 model.

Note that the Cyclops lens is described as an optional extra. It was not a standard feature until 1961. Most watches were supplied with an acrylic crystal with a cyclops lens fitted, but the purchaser could opt to change it for one without.

HOW TO USE YOUR GMT-MASTER

THE RED 24-HOUR HAND:
The 24 hour hand revolves once round the watch face every 24 hours. It is directly connected to the normal hour and minute hands and moves with them, either during the normal running of the watch or when the hands are being set.

It is, therefore, essential that the revolving rim be reset so that the triangle falls on 12 o'clock before each operation.

EXAMPLE OF USE:
Supposing you are leaving New York for Rome. The time shown on your watch is 14.19 hrs. You know that Rome time is six hours later than that of New York (see table below). Starting with the triangle at 12 o'clock (see paragraph above) you turn the revolving rim in an *anti-clockwise* direction through six hours, i. e. until the white triangle falls against the 9 o'clock position (three hours on the normal dial corresponding to six on the outer rim). You can now read both New York time (by the normal hour hand) and Rome time (by the red 24 hour hand) clearly and simultaneously. The minute and second hands apply, of course, to both places.

Conversely, when flying in the opposite direction (East–West) the outer rim is turned through the required number of hours but in a *clockwise* direction.

Remember : When travelling West–East turn anti-clockwise
When travelling East–West turn clockwise.

TIME ON THE EARTH'S SURFACE :

Alaska	03.00 hrs.
Los Angeles, San Francisco, Vancouver	04.00 hrs.
Denver, Phoenix	05.00 hrs.
Chicago, Mexico City, New Orleans	06.00 hrs.
Miami, New York, Washington, Montreal	07.00 hrs.
Buenos Aires, Santiago, Newfoundland	08.00 hrs.
Rio de Janeiro, São Paulo	09.00 hrs.
Azores	10.00 hrs.
G. M. T., London, Madrid	12.00 hrs.
Berlin, Geneva, Paris, Rome, Stockholm	13.00 hrs.
Alexandria, Cairo, Istanbul	14.00 hrs.
Tel Aviv, Moscow	15.00 hrs.
Karachi	17.00 hrs.
Calcutta	18.00 hrs.
Bangkok, Saigon, Singapore	19.00 hrs.
Hong Kong, Shanghai	20.00 hrs.
Darwin, Tokyo	21.00 hrs.
Sydney, Melbourne	22.00 hrs.
Christchurch, Wellington	24.00 hrs.

Cities on daylight saving time, add one hour

Ref. 6542

New York time : 14.19 hrs. indicated on dial. Rome time : 20.19 hrs. indicated on outer rim by 24-hr. hand.

THE ROLEX WATCH COMPANY LIMITED, GENEVA, SWITZERLAND
(H. WILSDORF, FOUNDER & CHAIRMAN)

LONDON, PARIS, NEW YORK, TORONTO, COLOGNE, BRUSSELS, DUBLIN, MILAN, ADEN, BUENOS AIRES, JOHANNESBURG, BOMBAY, MEXICO CITY, HAVANA, SINGAPORE, TOKYO, SAO PAULO.

© 1960 Montres Rolex S. A. RC. 260. III 60. 15 M. Printed by Haefeli & Co. Switzerland

---

THE GMT-MASTER... "THE FINEST TIMEPIECE FOR FLIGHT NAVIGATION"

(Major F. Libby, USAFR)

The GMT-MASTER was created by Rolex to meet the very special and exacting needs of the flying personnel of two world renowned aviation companies. Its success was immediate, 20 out of 21 aircraft captains declaring that the revolving rim and twenty-four hour hand constitute an important aid to airline operations.

Major Frederick A. Libby, an active airline navigator and a senior navigator in the USAF Reserve, who has for the past two years been continually using a GMT-MASTER, the chronometer which tells the time *clearly and simultaneously* in any two places on the earth's surface, says — "Two years of use has convinced me that the GMT-MASTER is the finest timepiece available to the flight navigator".

The GMT-MASTER has also met with an enthusiastic reception from ships' captains, members of the armed forces and international business men, in fact everybody to whom the knowledge of local time only is insufficient. With the GMT-MASTER Rolex have brought chronometric precision to the art of telling the time thousands of miles away.

The GMT-MASTER is equipped with a revolutionary revolving rim, calibrated into 24 hours, and a red 24-hour hand and, consequently, it will tell you the exact time at any two places on the earth at once. A pilot or navigator can now know, instantaneously and at any moment, Greenwich Mean Time as well as the local time wherever he may be.

The GMT-MASTER is an Officially Certified Chronometer which has passed the rigorous tests of a Swiss Institute for Official Timekeeping Tests. It is 100% waterproof in its Oyster case, and thus proof against dust and dirt as well. It is selfwound by the rotor Perpetual mechanism which, while it is worn, not only gives the vital assurance that the watch is fully wound but, by keeping the mainspring at a constant tension, also makes for greater accuracy. The date, shown in a neat dial-window and magnified by the 'Cyclops' lens, changes automatically and instantaneously each midnight.

On top of this, the revolving rim of the GMT-MASTER will serve innumerable purposes. It can be set to measure elapsed time, to check the duration of telephone calls, to time car speeds or it will remind you of an appointment. There is no end to the uses you will find for it.

REGISTERED AND PATENTED DESIGN, ALL COUNTRIES

CHECK THE ADVANTAGES OF THE GMT-MASTER

√ Revolutionary red 24-hour hand and revolving rim calibrated into 24 hours.

√ Simultaneous indication of local time for any two time-zones.

√ Rolex Oyster Case, guaranteed waterproof and pressureproof to 165 ft. (50 metres) under water.

√ "Twinlock" double-safety winding crown which *screws down onto the case.*

√ Unbreakable, crackleproof crystal.

√ Rotor Perpetual selfwinding mechanism, suspension-sprung against shock.

√ Date magnified by "Cyclops" lens (optional), changes automatically and *instantaneously* each midnight.

Actual size

Ref. 6542

Enlarged

A compass, too!
In the Northern hemisphere the Rolex GMT-Master may also be used as a compass. Simply point the hour-hand towards the sun and automatically the red 24-hour hand will point to the North. Try it...!

√ Officially Certified Chronometer movement, 25 jewels.

√ Shock-resisting and vibrationproof.

√ Guaranteed unbreakable mainspring.

√ Impossible to overwind.

√ Antimagnetic escapement.

√ Guaranteed "radiation-safe".

√ Stainless steel "Eversure" bracelet, registered "Flushfit" design.

√ The Rolex Red Seal is your guarantee that the GMT-MASTER to which it is attached has been submitted to a Swiss Institute for Official Timekeeping Tests and has been awarded an Official Timing Certificate with the proud title of *Chronometer*.

REGISTERED AND PATENTED DESIGN, ALL COUNTRIES

THE GMT-MASTER... "THE FINEST TIMEPIECE FOR FLIGHT NAVIGATION"

(Colonel F. Libby, USAFR)

The GMT-MASTER was created by Rolex to meet the very special and exacting needs of the flying personnel of two world renowned aviation companies. Its success was immediate, 20 out of 21 aircraft captains declaring that the revolving rim and twenty-four hour hand constitute an important aid to airline operations.

Colonel Frederick A. Libby, an active airline navigator and a senior navigator in the USAF Reserve, who has for several years past been continuously using a GMT-MASTER, the chronometer which tells the time *clearly and simultaneously* in any two places on the earth's surface, says — "Years of use has convinced me that the GMT-MASTER is the finest timepiece available to the flight navigator".

The GMT-MASTER has also met with an enthusiastic reception from ships' captains, members of the armed forces and international business men, in fact everybody to whom the knowledge of local time only is insufficient. With the GMT-MASTER Rolex have brought chronometric precision to the art of telling the time thousands of miles away.

The GMT-MASTER is equipped with a revolutionary revolving rim, calibrated into 24 hours, and a red 24-hour hand and, consequently, it will tell you the exact time at any two places on the earth at once. A pilot or navigator can now know, instantaneously and at any moment, Greenwich Mean Time as well as the local time wherever he may be.

The GMT-MASTER is an Officially Certified Chronometer which has passed the rigorous tests of a Swiss Institute for Official Chronometer Tests. It is 100% waterproof in its Oyster case, and thus proof against dust and dirt as well. It is selfwound by the rotor Perpetual mechanism which, while it is worn, not only gives the vital assurance that the watch is fully wound but, by keeping the mainspring at a constant tension, also makes for greater accuracy. The date, shown in a neat dial-window and magnified by the 'Cyclops' lens, changes automatically and instantaneously each midnight.

On top of this, the revolving rim of the GMT-MASTER will serve innumerable purposes. It can be set to measure elapsed time, to check the duration of telephone calls, to time car speeds or it will remind you of an appointment. There is no end to the uses you will find for it.

REGISTERED AND PATENTED DESIGN, ALL COUNTRIES

CHECK THE ADVANTAGES OF THE GMT-MASTER

√ Revolutionary red 24-hour hand and revolving rim calibrated into 24 hours.
√ Simultaneous indication of local time for any two time-zones.
√ Rolex Oyster Case, guaranteed waterproof and pressureproof to 165 ft. (50 metres) under water.
√ "Twinlock" double-safety winding crown which *screws down onto the case* and protected by shoulders.
√ Unbreakable, crackleproof crystal.
√ Rotor Perpetual selfwinding mechanism, suspension-sprung against shock.
√ Date magnified by "Cyclops" lens, changes automatically and *instantaneously* each midnight.

Ref. 1675   Actual size

A compass, too!
In the Northern hemisphere the Rolex GMT-Master may also be used as a compass. Simply point the hour-hand towards the sun and automatically the red 24-hour hand will point to the North. Try it...! In the Southern hemisphere it will point to the South.

Enlarged

√ Officially Certified Chronometer movement, 26 jewels.
√ Shock-resisting and vibrationproof.
√ Guaranteed unbreakable mainspring.
√ Impossible to overwind.
√ Antimagnetic escapement.
√ Stainless steel "Eversure" bracelet, registered "Flushfit" design.
√ The Rolex Red Seal is your guarantee that the GMT-MASTER to which it is attached has been submitted to a Swiss Institute for Official Chronometer Tests and has been awarded an Official Timing Certificate with the mention "Especially Good Results"

REGISTERED AND PATENTED DESIGN, ALL COUNTRIES

GMT-Master owner's booklet
English/North American version, 1965.

This later brochure features the GMT-Master, model 1675, which was launched in 1962. This model introduced the protective crown guards to the GMT-Master range.

Note that the Cyclops lens is now a standard feature.

Submariner owner's booklet
English/North American version, 1962.

Brochure for the Submariner, model 5512.

In 1961 the 6536, 6538 and 5510 models had ceased production. The 5508 model is mentioned as the only non-chronometer version of the Submariner available.

In late 1962, the non-chronometer Submariner, model 5513, was launched, and the 5508 model was gradually phased out.

The Rolex Submariner

Time in the twilight of deep water

Everyone who has had the chance to explore the splendours of the new world under the sea will know that the thrilling new sports, deep-sea diving, fishing or exploring, are bound to become increasingly popular as continually new equipment is designed and perfected.

One of the most essential requirements is a truly reliable watch, so completely waterproof that it will function perfectly in the great pressures of deep water. Those using Aqualung equipment need a watch that will time their fresh air supply, that will tell at a glance exactly how long they have been under water and how long they may stay there.

To meet these requirements Rolex are proud to offer the SUBMARINER.

Here is a watch that is guaranteed WATERPROOF and PRESSURE-PROOF to a depth of 660 ft. below water.

It is equipped with an external revolving bezel, surrounding the dial. This bezel is calibrated in fives from zero to sixty. The zero mark can be set against the second, minute, or hour hand and after any given period, the time elapsed can be read at a glance on the calibrated bezel.

The SUBMARINER is a specially designed adaptation of the world famous Rolex Oyster Perpetual.

It is fully automatic, motivated by the Rolex patented rotor Perpetual selfwinding mechanism.

It has been made super shock-resisting to withstand the hazards of diving, sailing and boating.

The mainspring is unbreakable.

The SUBMARINER is a strong and reliable wrist-watch the importance of which has already been recognized by those concerned in sea-going or sub-marine activities.

A deep-sea diver is searching for a sunken wreck. He has 35 minutes fresh air supply. Commencing his dive at 16.40 hrs., he sets the calibrated dial of his SUBMARINER to memorise that time for him.

He observes from his SUBMARINER that his descent has taken three minutes. Allowing five minutes for the ascent, he knows he must leave off his search when the minute hand reaches 30 on the calibrated bezel i. e. at 17.10 hrs.

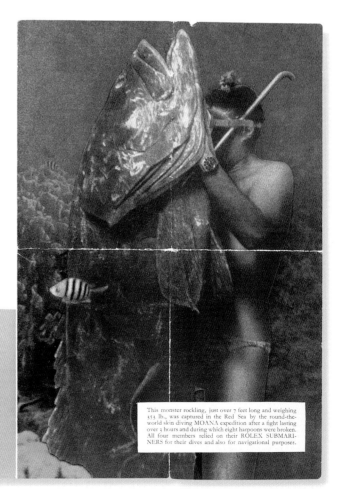

This monster rockling, just over 7 feet long and weighing 353 lb., was captured in the Red Sea by the round-the-world skin diving MOANA expedition after a fight lasting over 3 hours and during which eight harpoons were broken. All four members relied on their ROLEX SUBMARINERS for their dives and also for navigational purposes.

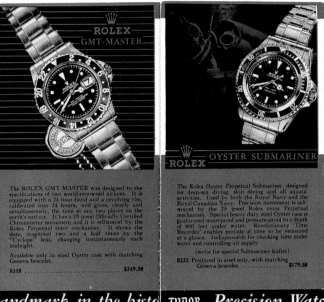

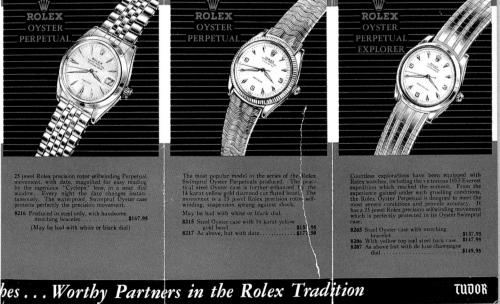

Rolex catalog
Canadian version,1968.

Left, the GMT-Master and Submariner models.

Below left are the three Explorer variants that sold during the 1950s and 1960s, in the North American market.

See page 72 for further information.

Rolex Milgauss

Milgauss owner's booklet
English/North American
version, late 1960s.

The Milgauss, model 1019, is
featured. This model replaced the
previous 6541 model.
 Two versions of the 1019
model are shown. The first has
luminous hands and luminous
paint on the baton hour markers.
 On the second version, the
luminous paint is replaced by
black.

150

The Rolex Milgauss
is an officially certified
chronometer which is both
waterproof (Oyster) and
selfwinding (Perpetual).
It also has a *high resistance to
magnetism.*

Cover illustration:
Photo of 28 GeV synchrotron,
CERN, Geneva.

The Rolex Milgauss

One of the consequences of present-day technological
development is that an ever increasing number of scientists
and technicians is exposed to magnetic fields of varying
intensities in the course of their day-to-day work. These
specialists have something in common: their watches must
keep excellent time.
One of the basic difficulties which they have to face is
that an ordinary watch placed in a magnetic field just over
60 Oersted begins to behave erratically and its movement
will stop working well before the magnetic force reaches a
strength of 1000 Oersted.
Rolex accepted this challenge and now, after several years
of research into the problem, it has produced a solution in
the form of the amazing Milgauss watch.
Independent tests, conducted by Rolex, have been double-
checked by well known laboratories in the USA, Great

ROLEX

Britain, Canada, Japan, the Netherlands and Switzerland
– which have all proved the qualities of this watch to their
own satisfaction. Here is one example – The European
Centre for Nuclear Research in Geneva (CERN) states:
"We may say with confidence that these watches appear
to us to be eminently suitable for wear in magnetic fields
up to 1000 Oersted and their accuracy in these conditions
is comparable to that of high quality watches which have
not been exposed to magnetic fields."
It can therefore be stated that the Rolex Milgauss is
virtually unaffected by magnetism. The limit at which a
normal watch stops working is below 100 Oersted, while
in the case of the Rolex Milgauss this limit has been
pushed beyond 1000 Oersted. Indeed, in experimental tests
of short duration, watches have been taken to 5000 Oersted

ROLEX

without any noteworthy effect of residual magnetism.
When exposed to a strong magnetic field, the Milgauss
chronometer shows slight variations as long as the experi-
ment lasts. However – and this is important – the residual
magnetism is of little or no consequence. Even after being
passed through a magnetic field of 1000 Oersted the
chronometer resumes its regularity of operation.
These astonishing results are attributable to two principal
factors:

— The use of anti-magnetic alloys, themselves the product
 of extensive research, to engineer certain parts of the
 escapement mechanism.

— Completely encapsulating the movement inside a
 continuous sheath of pure magnetic iron.

A Rolex Red Seal Chronometer

The Milgauss holds the jealously guarded *official* title of *Swiss Chronometer*. This means that the watch has successfully undergone all the tests of accuracy carried out by one of the seven Swiss Institutes for Official Chronometer Tests. These last 15 days and nights and include careful scrutiny of performance in five positions and at three specified temperatures.

After these severe trials, the movement and the case are subjected to a last and rigorous examination in our own laboratories before the watch is found worthy to bear the name of Rolex Oyster Perpetual Red Seal Chronometer. The confirmation is to be found in the words "Superlative Chronometer Officially Certified" which appear on its dial and in the Rolex Red Seal which accompanies the Milgauss.

The Rolex Milgauss

is an officially certified Rolex Oyster Perpetual Chronometer. It is selfwinding and guaranteed waterproof to a depth of 50 m (165 ft). *It also has a high resistance to magnetism.*

ROLEX

The Twinlock winding Crown
The Twinlock winder, with double safety device, screws down onto the case. It is waterproof and ensures perfect sturdiness at the weakest point of a watch.

The Oyster Bracelet
The typical Oyster bracelet fits the Oyster case so perfectly that it practically becomes part of it. It complements and enhances the Rolex Oyster.

The Perpetual Rotor
The principle of the Perpetual rotor mechanism – invented, improved and patented by Rolex – was later adopted by many other manufacturers of modern watches. Worn six to eight hours a day, a Rolex Oyster Perpetual will run for 48 hours.

ROLEX

Gauss/Oersted
The magnetic induction is expressed in Gauss.
The magnetic field is expressed in Oersted.

Weiss Domains
Weiss domains are groups of molecules of ferro-magnetic material in which all the magnetic moments are aligned in the same direction.

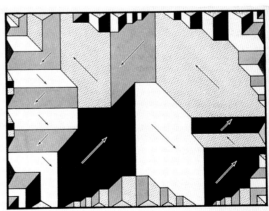

Weiss Domains
Reprinted from JOURNAL OF APPLIED PHYSICS, Vol. 36 No. 5 (1965, p. 1649) by courtesy of the American Institute of Physics.

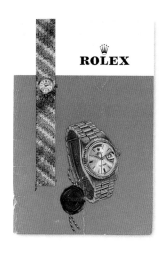

Rolex catalog
English version, 1966.

The watches are listed by catalog number rather than model number.

Listed below are sports models and their 1966 prices:

GMT-Master
Model 1675, catalog number 1015.
Price: £84 11 shillings/$126.

Submariner
Model 5513, catalog number 1029.
Price: £70 8 shillings/$105.

Explorer
Model 5504, catalog number 1022.
Price: £51 3 shillings/$76.

Below, the GMT-Master, model 1675, is fitted with the early pointed crown guards.

Datejust
1017 Rolex Oyster Perpetual Datejust (*above*). Stainless steel. Waterproof and self-winding Shows the date automatically, magnified by the "Cyclops" lens. With bracelet as shown, **£89 2s.** With leather strap, **£80 0s.**

Chronometer
1014 Rolex Oyster Perpetual Chronometer (*below*). Stainless steel. Waterproof and self-winding. With "Flush-fit" bracelet, **£72 18s.** With leather strap, **£68 7s.** 1027 As 1014, but with smooth bezel. With leather strap, **£65 1/s.**

ROLEX
OYSTER PERPETUAL
CHRONOMETER

Chronometer
3019 Rolex Oyster Perpetual Chronometer (*above*). 9ct. gold. Waterproof and self-winding. With 9ct. gold "Flush-fit" bracelet as shown, **£172 3s.** With leather strap, **£122 11s.**

3124 As 3019, but with the date shown automatically, magnified by the "Cyclops" lens. With 9ct. gold "Flush-fit" bracelet, **£181 5s.** With leather strap, **£131 13s.**

4

G.M.T. Master
1015 Rolex Oyster Perpetual Chronometer **G.M.T. Master.** Designed by Rolex to meet the special and exacting needs of the flying personnel of two world-renowned aviation companies. It has been acclaimed by pilots, navigators and by businessmen. The G.M.T. Master tells the time at any two places on the earth at once. The date is shown automatically, magnified by the "Cyclops" lens. Stainless steel. With "Flush-fit" bracelet, **£84 11s.**

AT YOUR SERVICE booklet, given with every Rolex and Tudor watch, shows how the G.M.T. Master may be simply set to show, as an example, London time and date and New York time, and explains the ease with which it may be read at a glance. Pan American Airways photograph.

THE
MEANING OF
CHRONOMETER

Every Rolex Oyster Perpetual Chronometer has been awarded an Official Timing Certificate by the independent, impartial Swiss Institutes for Official Chronometer Tests. To receive this award, the watch movement must successfully undergo a series of rigorous tests which are carried out during fifteen days and nights, in five different positions and at three different temperatures.

Over the years, less than four watches in every thousand produced in Switzerland have obtained these Official Timing Certificates. A far higher proportion of the production of Rolex watches for men win Official Chronometer Certificates than that of any other Swiss watch manufacturer.

Now the Rolex standards are even higher. No Rolex is called Chronometer and has the Red Seal attached unless it has obtained the very highest distinction for precision and quality from a Swiss Institute for Official Chronometer Tests: "Especially Good Results".

5

Below, on the left, the Submariner, model 5513. This particular example is fitted with the optional Explorer-style dial.

Below, on the right, the Explorer, model 5504. The dial is marked *Super Precision.*
   Both the Explorer 5504 and 5500 models were signed *Super Precision* on the dial. The 5504 may be distinguished from the smaller 5500 model by its larger winding crown.

THE

PERPETUAL

MOVEMENT

The chief advantage of a self-winding watch may seem to be that the wearer does not have to remember to wind it. To the watchmaker it means more—it means greater accuracy.

The Rolex Perpetual winds itself with every movement of the wrist. The rotor swings into different positions as the watch is moved. It is silent, able to revolve in either direction, smooth in action, yet permanently connected to the mainspring. Once the mainspring is fully wound, an ingenious slipping clutch mechanism prevents overwinding. The result—the mainspring is always at the right tension, giving far greater accuracy.

Rolex patented the rotor self-winding mechanism in 1931, an invention so successful that the basic principle was adapted by the entire watch industry when the patents expired. But fifteen years start, and constant research and improvement keeps the Rolex Perpetual in the forefront to this day.

6

Submariner

1029 Rolex Oyster Perpetual Submariner. The strengthened stainless steel Oyster case is designed to withstand water pressure to a depth of 660 feet. The Submariner is the finest watch for divers, but it is equally suitable for everyday wear, especially for those who require accurate timekeeping in hard or unusual conditions—medical or industrial research, surveyors, or oil, mining and civil engineers. The calibrated bezel can be set to act as a reminder of a given time. With "Flush-fit" bracelet, £70 8s.

The Submariner is officially used by divers of the Royal Navy, the French Navy, the Royal Canadian Navy, the Royal Australian Navy. From the experience gained from innumerable expeditions, Rolex have developed watches to meet important, exacting requirements. Admiralty photograph, Crown copyright.

ROLEX
OYSTER PERPETUAL

Air-King

1023 Rolex Oyster Perpetual Air-King (*above*). Stainless steel. Waterproof and self-winding. With leather strap, **£44 11s.** With "Flush-fit" bracelet, **£49 2s.**

3123 As 1023, but 9ct. gold case. With leather strap, **£98 15s.** With 9ct. gold "Flush-fit" bracelet, **£148 7s.**

1030 Rolex Oyster Perpetual Air-King-Date (not illustrated). As 1023, but shows the date automatically, magnified by the "Cyclops" lens. With leather strap, **£53 14s.** With "Flush-fit" bracelet, **£58 5s.**

7

2009 Rolex Oyster Perpetual for ladies. Stainless steel. Engine turned bezel. With leather strap, £63 16s. With "Flush-fit" bracelet, £68 7s.

2007 Similar to 2009, but smooth bezel and automatic date, magnified by the "Cyclops" lens. With leather strap, £66 17s.

4081 Similar to 2009, but 9ct. gold, £99 15s.

4074 Similar to 2009, but 18ct. gold, £125 11s.

Explorer

1022 Rolex Oyster Perpetual Explorer (*below*). Has been used in the Arctic, the Himalayas and the tropics. Designed to give accuracy under the toughest out-of-door conditions. Stainless steel. With "Flush-fit" bracelet, £51 3s.

ROLEX

- the one with
substantial
character

Rolex Oyster catalog
English/North American
version, 1966.

Brochure showing the
Submariner, model 5513, with an
optional Explorer-style dial; the
Explorer, model 1016 and the
GMT-Master, model 1675.
   The Submariner model is
shown with a picture of a British
Royal Navy diver wearing an
earlier Submariner, model 6538.

Sir John Hunt, who led the
celebrated Everest expedition of
1953, is featured to promote the
Explorer, and a photograph of Pan-
American Airline staff is used to
promote the GMT-Master.
   Rolex catalog numbers are
positioned below the watches.

*Rolex watches are not
"styled"—they are designed. With
purpose. Consider these facts:*
Rolex invented the waterproof wrist-watch.
The Oyster case, entirely Geneva-made, is
still the only one of its kind in the world. It
takes 162 separate precision operations to
create the Oyster case from a solid block of
fine stainless steel or gold. It is not moulded,
pressed from sheet metal or soldered; it has
immense strength. The Oyster case alone
costs as much to make as most other
complete watches.
Rolex made the first rotor self-winding
wrist-watch—so successful that all other
automatic watch manufacturers just had to
adopt the Rolex principle. Rolex, with
constant research and development, keeps
ahead.
Rolex Oyster watches have been official
equipment on rugged Himalayan expeditions,
and have been down to the bottom of the
deepest ocean chasm. They have never been
affected by heat and cold, or water, or sand,
or dust.
Rolex make but a very small percentage of
the 50 million watches made in Switzerland
every year. Yet Rolex have been awarded
nearly half of the Swiss Official Chronometer
Certificates during the last 40 years. A
chronometer takes nearly a year to make.
It must undergo a series of stringent tests
and pass with the mention "Especially Good
Results."
When a man has a world in his hands, you
find a Rolex on his wrist.

Rolex Oyster Perpetual Submariner
Created originally for deep-sea divers, the Submariner
is nearly half an inch thick and weighs a hefty 3 ounces.
It's guaranteed waterproof down to 660 feet underwater
and is self-winding. The calibrated auto-lock
turning bezel permits easy and precise calculation of
decompression times and also the length of telephone
conversations, photo development, etc.
The Submariner has a wealth of other features (and
recently, a number of imitators). It has all the character
of an original. Officially supplied to the Admiralty for
use by Royal Navy divers.

1029
Stainless steel with "Flush-fit" bracelet   £70 8s.

Rolex Oyster Perpetual Explorer
Has been used in the Arctic, the tropics and the
Himalayas. Supplied to the first successful Everest
expedition. In the words of Sir John Hunt, leader of
the Expedition, who is seen below: "We have come to
look upon Rolex watches as an important part of
high-climbing equipment."
Officially Certified Chronometer with the special
mention "Especially Good Results." Self-winding.

1025
Stainless steel with "Flush-fit" bracelet   £72 18s.

Rolex Oyster Perpetual Datejust
One of the watches which have made Rolex world
famous. The Datejust tells the time and shows the
date, which changes automatically at midnight, in a
neat window on the dial, magnified by the Cyclops
lens for easy reading. Officially Certified Chronometer
which has been awarded the special mention
"Especially Good Results." Waterproof. Self-wound
by the Rolex Perpetual rotor mechanism.

3007
18ct. gold with 18ct. gold bracelet   £411 2s.
1017
*Stainless steel with two-tone finish stainless steel
"Flush-fit" bracelet*   £89 2s.
With leather strap   £86 0s.

Rolex Oyster Perpetual GMT-Master
Originally created for airline pilots, ships' captains and
navigators, the GMT-Master has been adopted by
international travellers, sportsmen and successful
businessmen the world over. With an auto-lock
revolving bezel and 24-hour hand, it shows clearly and
simultaneously the time in any two time zones. Like
every Rolex Oyster, the waterproof case as well as its
rotor self-winding movement is entirely Swiss-made to
the most exacting standards of precision. Officially
Certified Chronometer with the mention "Especially
Good Results." The date is shown automatically,
magnified by the Cyclops lens.

1015
Stainless steel with "Flush-fit" bracelet   £84 11s.

Rolex Oyster Perpetual Day-date
The crowning achievement of Rolex and of the Swiss
watchmaking industry. The Oyster case (waterproof,
of course) is hewn from solid 18ct. gold or platinum
to the famous Oyster shape by highly skilled Genevan
craftsmen. The Perpetual rotor self-winding movement
has achieved the highest distinction that can be
awarded by the Swiss Institutes for Official
Chronometer Tests. Shows the date, magnified by the
Cyclops lens, and the day spelt in full. This is the
Rolex, possibly the most brilliant timepiece today,
which is worn by men who guide the destinies of the
world.

3131 *Solid platinum with solid platinum matching bracelet.
The dial is surrounded by 46 fine diamonds, and the hours are
marked by ten carefully chosen diamonds*   £1,975 0s.
3035 *18ct. gold with magnificent 18ct. gold bracelet*
£459 14s.
With leather strap, 18ct. gold buckle   £243 0s.

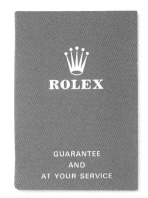

GUARANTEE
AND
AT YOUR SERVICE

Put briefly, once the Rolex Day-date has been set correctly to show the day and the date, all that has to be done on the morning following a month of less than 31 days is to turn the hands back to 6 o'clock of the previous evening and then to advance them again.

On the first day of March this operation carried out three times (twice in a leap year) will correct the date.

The Rolex GMT Master

The Rolex GMT Master has the normal second, minute and hour hands and the date, and in addition a revolving bezel, calibrated into 24 hours and a red 24 hour hand. Consequently it may be used to tell the exact time at any two places on the earth at once.

The red 24 hour hand revolves once around the dial every 24 hours and is directly connected to the normal hour and minute hands and moves with them, either during the normal running of the watch or when the hands are being set.

24

The red 24 hours hand, with the triangle tip, is read on the revolving bezel outside the dial

The GMT Master is easy to read if it is remembered that the normal hour hand is read on the dial and the red 24 hour hand is read on the revolving bezel outside the dial.

Suppose it is 2 o'clock in the afternoon Greenwich Mean Time and it is required to set the GMT Master to GMT and date and New York time.

First set GMT and date, using the normal hands as described already under "Setting the Rolex Oyster" (page 14), "Setting to the second" and "Setting the Date" (pages 20 and 21).

When this has been done, and with the red 24 hour triangle of the revolving bezel in the position corresponding to 12 o'clock on the dial, the red

25

Rolex guarantee brochure
English version, 1968.

This type of booklet came with purchase during the 1960s and 1970s. The individual case and model number, along with the date of purchase, were entered onto the brochure's front page.

The book contains an explanation of the servicing process, along with descriptions of other Rolex models. The GMT-Master, model 1675, is shown.

155

hand will read 14.00 hours—it, too, shows GMT. Knowing that New York is five hours behind GMT, the bezel must be turned five hours in a clockwise direction. A downward pressure on the bezel will enable it to be turned to the appropriate position, after which it will lock. The red hand now reads 09.00 hours on the bezel—New York time. The minute and seconds hands apply, of course, to both places.

For places west of Greenwich, the bezel must be turned in a clockwise direction. For places east of Greenwich, in an anti-clockwise direction.

Thus, having set the GMT Master to GMT or other "home" time, and date, on the dial as for a normal date watch, it is a simple matter to set the bezel to the other required time, knowing the number of hours it is ahead or behind GMT or "home" time.

The Rolex Submariner

The Rolex "Submariner" has a rotating bezel which provides a simple means of recording elapsed time. This bezel is scaled to 60, which is marked by a prominent triangle. If this 60 triangle is turned until it is opposite the minute hand at the

26

commencement of a dive or an operation, the time elapsed since the start can be read at a glance on the revolving bezel. A downward pressure on the bezel will enable it to be turned to the appropriate position, after which it will lock.

Setting the hands of the "Submariner" is as described under "Setting the hands of an Oyster watch" and "Setting to the second".

The Tudor Advisor

The Tudor Advisor is an alarm watch. There are two winding buttons, one at the 2 o'clock position and the other at the 4 o'clock position.

The button at the 4 o'clock position is for winding and setting the hands, just as a normal watch.

The button at the 2 o'clock position controls the alarm mechanism. In its normal position it is used to wind the alarm. To set the alarm, pull out the button and turn in an anti-clockwise direction. To have the alarm set for functioning this button must be in the hand-setting position.

Pushing in the button at the 2 o'clock position stops the alarm.

27

The Rolex 'Submariner' is the official watch of divers of many of the world's navies

The Rolex 'Cosmograph' . . . precision wrist-watch, stop-watch, tachometer

We invented this for hunting treasure 660 feet under the sea

We invented this to track the speed of a 7-litre sports racer

This one has a bit of both and a substantial character of its own

Like the 'Submariner' and the 'Cosmograph,' the Rolex Oyster Perpetual 'Datejust' is not for every man. But its *character* may suit you.

The 'Datejust' combines the ruggedness of the 'Submariner' with the versatility of the 'Cosmograph.' And it has distinctive features of its own.

The Geneva-made Oyster case is hewn from a solid block of Swedish steel or gold. The bracelet, finished by hand with the same precision as the case and movement, is designed specially for the case. The movement has won the highest distinction for precision and quality a Chronometer can normally obtain. The Perpetual Rotor system keeps the watch wound at a constant, ideal tension. A calendar shows the date magnified by the 'Cyclops' lens.

The 'Datejust' is not "styled," it is designed. Honestly. With purpose. Wear it and you can dive for treasure off the coast of Nassau, shave seconds off the lap record at Monza, or address the United Nations.

With a Rolex on your wrist, you have entire worlds in your hands.

When a man has a world in his hands, you expect to find a Rolex on his wrist **ROLEX** GENEVA

Aden Auckland Bandung Bangkok Bombay Brussels Buenos Aires Cologne Dublin Havana Hong Kong Johannesburg London Madrid Melbourne Mexico City Milan New York Paris São Paulo Singapore Sydney Tokyo Toronto

Playboy
Magazine advertisement, 1966.

The sports models featured are the Submariner, model 5513, fitted with the optional Explorer-style dial, and the Cosmograph 6239 model.

156

Explorer owner's booklet
English/North American
version, 1969.

Brochure for the Explorer, model
1016. It includes testimonials
supporting the watch's
toughness and reliability.

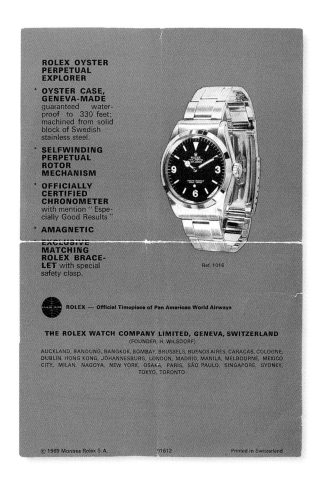

**ROLEX OYSTER
PERPETUAL
EXPLORER**

* **OYSTER CASE,
GENEVA-MADE**
guaranteed water-
proof to 330 feet;
machined from solid
block of Swedish
stainless steel.

* **SELFWINDING
PERPETUAL
ROTOR
MECHANISM**

* **OFFICIALLY
CERTIFIED
CHRONOMETER**
with mention "Espe-
cially Good Results"

* **AMAGNETIC**

* **EXCLUSIVE
MATCHING
ROLEX BRACE-
LET** with special
safety clasp.

Ref. 1016

ROLEX — Official Timepiece of Pan American World Airways

THE ROLEX WATCH COMPANY LIMITED, GENEVA, SWITZERLAND
(FOUNDER, H. WILSDORF)

AUCKLAND, BANDUNG, BANGKOK, BOMBAY, BRUSSELS, BUENOS AIRES, CARACAS, COLOGNE,
DUBLIN, HONG KONG, JOHANNESBURG, LONDON, MADRID, MANILA, MELBOURNE, MEXICO
CITY, MILAN, NAGOYA, NEW YORK, OSAKA, PARIS, SÃO PAULO, SINGAPORE, SYDNEY,
TOKYO, TORONTO

The Explorer, an Oyster Perpetual Chronometer, was created specially for the first successful Everest expedition. The specification was simple yet exacting – to design a thoroughly dependable, shockproof rugged watch to perform in the most difficult circumstances. After the expedition had successfully completed its mission, we received these comments :

BRIGADIER SIR JOHN HUNT

... As I have emphasized before, this expedition was built on the experience and achievement of others. Rolex Oyster watches have accompanied many pre-vious pioneering expeditions. They per-formed splendidly and we have indeed come to look upon Rolex Oysters as an important part of high climbing equip-ment...

SIR EDMUND HILLARY

... I wore the watch (Rolex Oyster Per-petual) continuously night and day... In the course of the expedition it experienced considerable extremes of temperature from the great heat of India to the cold temperatures at over 22,000 feet and seemed unaffected by the knocks it received on the rock climbs or the continual jarring of long spells of step cutting in ice... Its accuracy is all one could desire and it has run continuously ever since I put it on some nine months ago...

SHERPA TENSING

... your watches behaved splendidly under any and all the worst conditions imaginable, not only this time but also on previous expeditions to Mount Everest, of which I happened to have also been a member.

The Explorer has become recognized the world over as the watch for climbers, cave explorers, skiers :

BRITISH NORTH GREENLAND EXPEDITION

... occasional time signals broadcast from England proved that my Rolex watch was maintaining a remarkable accuracy. On no occasion did it require to be wound by hand. When on the ice-cap, away from base for several weeks at a time, it was of inestimable value to have on my wrist a watch whose accuracy could be relied upon at all times...

ARGENTINE ARMY EXPEDITION TO THE ANTARCTIC

An officer says :
As a pilot of a Cessna reconnaissance plane and later at the base of Sobral (81° S latitude and 40° W longitude) I had occasions to test out the qualities of my Rolex. It was exposed to temperatures of –53° C and to very rugged conditions. Its accuracy and reliability make it of incalculable value to all who have to work in completely adverse condi-tions, such as are found in the Antarctic.

BRITISH SPELEOLOGICAL EXPEDITION TO THE CANTA-BRIAN MOUNTAINS (SPAIN)

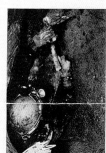

Caving is not a gentle activity and of necessity the Rolex Explorer watches received a great deal of rough treatment including immersion in mud and water, and scratching and banging against rock, but the accurate reading of time, which is an important safety factor in caving, never gave us any worries.
A. C. Huntington

ROYAL SOCIETY ANTARCTIC EXPEDITION

(My Rolex) ... became part of me – an unquestioned, reliable part – doing no more than gain a matter of seconds per week, although exposed to such extremes of temperature as a hot cooking stove, or digging out stores in minus 60° Fahrenheit.
Surgeon Lt-Cdr Dalgliesh, Leader Advance Party

The Explorer has a specially strengthened **pressure-proof** Oyster case which is waterproof and swimproof to 330 feet. It has been specially designed, with its officially certified chronometer movement, to be fully shockproof, to withstand the jars and knocks of mountain climbing, of exploring. It can safely be worn while operating a pneumatic drill or cutting rock-steps. It is selfwound by the Rolex Perpetual mechanism and the mainspring cannot overwind. Like all Rolex Oysters, the Explorer has the exclusive Twinlock screw-down winding button. The crown becomes an integral part of the case and is protected against accidental damage.

Not everyone is an explorer but there are many who are forced to treat their watch roughly. With the Explorer you can at last own a really accurate watch that is tough.

Rolex catalog
English version, 1969.

Below is the cover and the accompanying price list with the cost shown in pounds and shillings.
    The watches are listed by catalog number rather than model number.

Listed below are sports models and their 1969 prices:

Cosmograph
Model 6262, catalog number 1031.
Price: £109/$163.50.

Submariner
Model 5513, catalog number 1029.
Price: £84 10 shillings/$126.

GMT-Master
Model 1675, catalog number 1015.
Price: £101/$151.50.

Explorer
Model 1016, catalog number 1025.
Price: £87/$130.50.

Opposite page above, the Cosmograph, model 6262. On the left of the page is the standard dial, and to its right, the exotic dial, now known to collectors as the Paul Newman dial.

Opposite page below, the Submariner, model 5513. The text explains that the watch was being used by the British Royal Navy, French Navy and the navies of the Commonwealth.

Rolex Cosmograph

1031 Put the Cosmograph on your wrist. It is two watches in one: a highly accurate, masculine everyday wrist-watch and a precision stop-watch (or chronograph) always ready to measure elapsed time accurately to a fifth of a second. The Rolex Cosmograph was designed for the sportsman, the engineer in today's fast moving, time-conscious world. Rugged, reliable and simple to work. Protected from dust, dirt and perspiration by a patented Geneva-made case. There are minute and hour elapsed time recorder dials. The Cosmograph is available with white/black or black/white dial combinations. Stainless steel with stainless steel "Flush-fit" expanding bracelet.

For **amateur sailors** . . . it calculates speed, distance from the shore and position at night.
For **pilots**, **navigators** or **passengers of a small plane** . . . it checks ground speed and wind speed.
For **sportscar enthusiasts** . . . it calculates average speeds and lap-times at rallies or motor races.
For **athletes** in training . . . it records lap-times.

For **photographers** . . . it records camera shutter speeds and processing times.
For **production engineers**, **scientists**, **doctors**, **operations researchers**, **sales executives** . . . it times any operation or series of operations. Fractional timing is becoming more and more important and stage times, production rates, pulse beats, sales calls can all be measured quickly, easily, accurately.

3

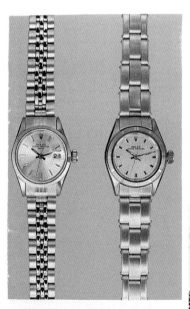

The Submariner is officially used by divers of the Royal Navy, the French Navy and Commonwealth Navies. It is the finest watch for divers, but is equally suitable for everyday wear, especially for those who require accurate timekeeping in hard or unusual conditions — medical or industrial research, surveyors, or oil, mining and civil engineers, for example.

Submariner

1029 Rolex Oyster Perpetual Submariner. Strengthened stainless steel Oyster case designed to withstand the pressure of water to a depth of 660 feet. The calibrated bezel can be set to act as a reminder of a given time. With "Flush-fit" expanding bracelet.

2007 (*above, left*) **Rolex Oyster Perpetual Ladydate.** Stainless steel. Shows the date automatically, magnified by the "Cyclops" lens. With "Flush-fit" bracelet as shown or with leather strap.

2009 (*above, right*) **Rolex Oyster Perpetual,** for ladies. Stainless steel. Engine turned bezel. With "Flush-fit" expanding bracelet or with leather strap.

7

Rolex catalog
English version, 1969.
continued

Below, the GMT-Master, model 1675. Its development is explained in this brochure, and because it was still the official timepiece of Pan-American Airlines, their endorsement is included.
Bottom, the Explorer, model 1016.

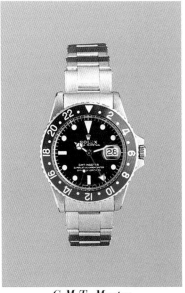

160

To merit the cherished title of Chronometer, each watch must be individually tested for fifteen days by a Swiss Institute for Official Chronometer Tests. The Rolex Red Seal signifies that the watch has been awarded the highest distinction for precision and quality, the special mention "especially good results".

Although Rolex makes just a tiny fraction of the total production of Swiss watches, Rolex have won nearly half of all the Chronometer certificates ever awarded.

G.M.T. Master
1015 Rolex Oyster Perpetual Chronometer G.M.T. Master. Stainless steel. With revolving outer bezel and 24-hour hand it tells the time at a glance in any two time zones. The date is shown automatically, magnified by the "Cyclops" lens. "Flush-fit" expanding bracelet.

8

The G.M.T. Master was designed by Rolex to meet the very special and exacting needs of the flying personnel of two world-renowned aviation companies. Its success was immediate, aircraft captains declaring that the revolving bezel, equipped with an automatic locking device to prevent accidental movement, and the twenty-four hour hand constitute an important aid to airline operation. It has also met with an enthusiastic reception from ships' captains, members of armed forces and international business men to whom the knowledge of local time only is insufficient.

The G.M.T. Master tells the time at any two places on the earth at once. It will, for example, show London time on the normal dial (and London date, too) and New York time with the red 24-hour hand on the calibrated outer bezel.

 Rolex — Official timepiece Pan American World Airways

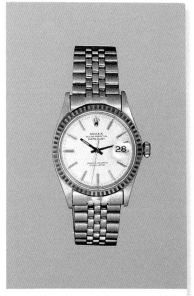

Explorer
1025 Rolex Oyster Perpetual Chronometer Explorer. Stainless steel, strengthened case designed to withstand water pressure to a depth of 330 feet. The Explorer has been used in the Arctic and Antarctic, the Himalayas and the tropics, and is designed to give accuracy under the toughest out-of-door conditions. With "Flush-fit" expanding bracelet.

Chronometer
1027 Rolex Oyster Perpetual Chronometer. Stainless steel. With "Flush-fit" expanding bracelet or leather strap.

9

Datejust
1017 Rolex Oyster Perpetual Chronometer Datejust. Stainless steel, with engraved bezel. The date is shown automatically, magnified by the "Cyclops" lens. With stainless steel "Flush-fit" bracelet as shown or leather strap.

Rolex catalog
English version, 1972.

Below is the cover and the accompanying price list with the cost shown in pounds.
The watches are listed by model number.

Listed below are sports models and their 1972 prices:

GMT-Master
Model 1675, in steel.
Price: £124/$186.

Cosmograph
Model 6265, in steel.
Price: £161/$241.

Submariner
Model 1680, in steel.
Price: £138/$207.

Sea-Dweller
Model 1665, only available in steel.
Price: £138/$207.

Explorer
Model 1016, only available in steel.
Price: £107/$160.

Milgauss
Model 1019, only available in steel.
Price: £124/$186.

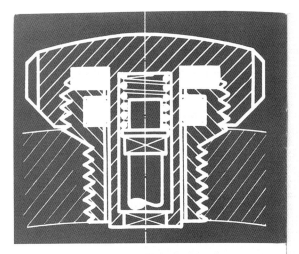

The distinctive Oyster case is hewn out of a solid block
of stainless steel or gold or platinum.
Such care and precision
are lavished on the Oyster case (162 separate operations
from block to final hand polishing)
that it alone costs more to make than most complete
watches. Each Rolex Oyster case is
guaranteed waterproof and pressure-proof*. Each Rolex
Oyster case is made entirely in Geneva.

*When case, crown and crystal are intact.

2

The key to this amazing
water-resistance is the
unique Rolex Twinlock
crown, fitted to all Rolex
Oysters. Like a submarine
hatch, the Twinlock
crown screws into the case.
When an experimental
Rolex Oyster was clamped on
the *outside* of Professor Jacques
Piccard's bathyscaph "Trieste",
submerged to a depth of seven
miles and subjected to a pressure
of *nearly 7 tons per square inch*, the watch
returned to the surface working perfectly.

3

Rolex watches helped Sir John Hunt conquer Everest.
Sir Francis Chichester depended on Rolex during his
solitary voyage around the world. Rolex watches have been
frozen in the Arctic, burned in the Sahara,
and punished almost to destruction under conditions
where no other watch could have survived.
It takes more than a year to make a Rolex Oyster.
Because so much of the work is done by
hand, production is limited to relatively small
numbers. Therefore, the Rolex Oyster
of your choice may not be immediately available.
In this age of mass production waiting
is more than compensated by the pride of knowing you
own one of the best watches in the world.
Remember that, as each Rolex is built to order, there are
no "standard" models. Most Rolex
watches are available with a variety of faces,
bezels, bracelets, and finishes.

8

When special
occupations make
special demands on a
watch, Rolex provides the
answer, like the Rolex "GMT-
Master" designed originally for the
exacting needs of flying personnel (official
timepiece of Pan American World Airways).

9

Rolex catalog
English version, 1972.
continued

Top, the text explains the Rolex
waterproofing system. The twinlock winding crown which was
fitted to the Submariner range is
shown. At this time a trip-lock
crown was also in existence, but
was only fitted to the Sea-
Dweller, model 1665.

Bottom, the brochure lists many
examples of the Rolex Oyster's
strength and dependability.

The right hand page shows an
early GMT-Master, model 1675.
Note the small triangle on the
secondary hour hand. This was
later replaced by a triangle almost
twice the size.

THE GMT-MASTER

THE COSMOGRAPH

THE SUBMARINER

THE SEA-DWELLER

1675

6265

1680

1665

The GMT-Master — created for pilots, navigators and world travellers. Rotor self-winding, officially certified chronometer, with date, guaranteed waterproof in its Oyster case. Revolving outer bezel and 24 hour hand tell the time simultaneously in any two time zones. Official watch of Pan American World Airways. Available in stainless steel or 18ct. gold.

The Cosmograph — a stop watch, guaranteed waterproof to a depth of 165 feet in its Oyster case, when push buttons and winding crown are screwed down. Created for sportsmen as well as for engineers, organisation and methods specialists, technicians, scientists and doctors.

The Submariner — the *original* deep sea diver's watch. Oyster case guaranteed waterproof and pressure-proof down to 660 feet. Rotor self-winding, officially certified chronometer with date. Revolving bezel measures elapsed diving time (of vital importance when diving). Official diver's watch of many of the world's navies. "Crown Guard" shoulders protect the winding crown. Available in stainless steel or in 18ct. gold.

The Sea-Dweller is a self-winding officially certified chronometer with date, and is guaranteed waterproof to a depth of 2,000 feet (610 metres). Fitted with a patent valve allowing helium which will even penetrate the watch glass when used in an underwater "habitat" to rapidly escape on decompression.

14

15

THE EXPLORER

THE MILGAUSS

THE DATEJUST

THE OYSTER PERPETUAL DATE

1016

1019

1603

1500

The Explorer — sometimes called the Hard Rolex. Specially strengthened Oyster case guaranteed waterproof to 330 feet. Super shock resisting, rotor self-winding, officially certified chronometer movement. The black face has extra luminous markings for easy reading under extreme conditions. The watch Sir John Hunt wore on his Everest expedition. Available in stainless steel only.

The Milgauss — most good watches are unaffected by magnetic fields of 60-70 gauss, but the Milgauss maintains superlative accuracy in magnetic fields of up to 1,000 gauss. Waterproof Oyster case, rotor self-winding, officially certified chronometer movement. In stainless steel with matching bracelet.

The Datejust in steel. Rotor self-winding, officially certified chronometer, with date changing automatically and instantaneously at midnight.

Rolex Oyster Perpetual Date — Rotor self-winding, officially certified chronometer movement. Oyster case guaranteed waterproof to 165 feet. The date changes automatically and instantaneously at midnight. Available in gold or in stainless steel with a variety of dials.

16

17

Above, the sports model range. Note that the GMT-Master's secondary hour hand features the larger triangle.

Rolex Cosmograph

Cosmograph owner's booklet
English/North American
version, early 1970s.

From the early 1970s, the
Cosmograph models 6263 and
6265 were the only two
Cosmographs available.
  The eighteen-carat gold
versions of the 6263 and 6265
shown here were tested to
chronometer standard but they
did not carry the words
*Superlative Chronometer Officially
Certified* on the dial. This wording
was introduced in the mid 1970s.

ROLEX

The Rolex Cosmograph

A stop watch, guaranteed
waterproof to a depth of
50 meters in its Oyster case,
with push buttons and Triplock
winding crown screwed down
onto the case.

Ref. 6265/0

The Rolex Oyster Cosmograph

Put the Rolex Oyster Cosmograph on your wrist and you
have two watches in one. A precise masculine instrument
which can measure elapsed time accurately to a fifth of a
second. The Cosmograph was created for sportsmen as
well as for engineers, organisation and methods specialists,
technicians, scientists and doctors.
Each of them recognises the need for precise timing,
whether it be the accurate measurement of a patient's
pulse, the length of an international telephone call, timing
a race, or checking the productive capacity of a machine.
And wearing this same versatile watch, he can dive to a
depth of 50 meters.

1

The Rolex Cosmograph

Available in steel with black
or engraved bezel; the face
is white with black dials or
vice versa.
It is also available in 18 ct
yellow gold with black or
engraved bezel; the face can be
gilded with black dials,
alternatively the face can be black
and the dials gilded.
Gold models are officially
certified Chronometers.

Ref. 6263/0

Ref. 6263/8   Ref. 6265/8

2   3

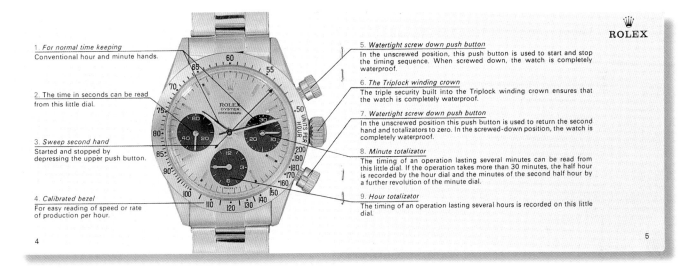

1. *For normal time keeping*
Conventional hour and minute hands.

2. The time in seconds can be read from this little dial.

3. *Sweep second hand*
Started and stopped by depressing the upper push button.

4. *Calibrated bezel*
For easy reading of speed or rate of production per hour.

5. *Watertight screw down push button*
In the unscrewed position, this push button is used to start and stop the timing sequence. When screwed down, the watch is completely waterproof.

6. *The Triplock winding crown*
The triple security built into the Triplock winding crown ensures that the watch is completely waterproof.

7. *Watertight screw down push button*
In the unscrewed position this push button is used to return the second hand and totalizators to zero. In the screwed-down position, the watch is completely waterproof.

8. *Minute totalizator*
The timing of an operation lasting several minutes can be read from this little dial. If the operation takes more than 30 minutes, the half hour is recorded by the hour dial and the minutes of the second half hour by a further revolution of the minute dial.

9. *Hour totalizator*
The timing of an operation lasting several hours is recorded on this little dial.

4

5

Various uses of the Rolex Cosmograph

In effect, the Rolex Cosmograph is two watches in one: it is a highly accurate stop watch recorder which, in the operating position is as resistant to dust and water as a normal watch; but it is also a Rolex Oyster and when the control buttons are screwed down it can be taken to a depth of 50 meters under water!

When the Cosmograph is used as a stop watch it can measure speed per hour, the production rates of machines, how long it takes to cover 100 meters or a circuit of the track, the length of an international telephone call and a thousand other things.

The Rolex Cosmograph has a tachometer or calibrated bezel, accurate time totalizators, a sweep second hand accurate to a $1/5$th of a second as well as all the properties of a normal watch. And it is protected by the amazing Oyster case.

How to operate the Cosmograph

To start, depress the upper push-button (in the unscrewed position). See page 5, point 5. As soon as a kilometer or a mile has been covered, depress again the same push-button. The sweep second hand then indicates the average speed on the engraved bezel in either miles or kilometers. To return the sweep second hand and the recorder dials to zero, depress now the lower push-button. See page 5, point 7.
To time the hourly production rate of a machine, select a unit of measurement: 100, 1,000 – whatever you like.
Start the sweep second hand and stop it when the number selected has been produced: for instance 100. The sweep second hand now indicates on the calibrated bezel the number of times this quantity can be produced in an hour. If the sweep second hand points to 60 when it is stopped, a simple calculation 60x100 reveals the hourly production rate: 6,000. If the sweep second hand points to 180, then the production rate of the machine is 180x100 = 18,000.

6

7

Rolex Oyster catalog
English version, 1973.

This brochure shows the entire Oyster range.

Below, the Explorer, model 1016, and Explorer II, model 1655, are shown. The seconds hand of the Explorer II does not yet display a luminous circle - this was not added until late 1974.

Rolex Oyster Perpetual 1035

Chronometer made in 18 carat gold only, "moiré" finish bezel with matching 18 carat gold "moiré" bracelet.

Rolex Oyster Perpetual 1013

Officially certified chronometer, selfwinding movement, guaranteed waterproof to 165 feet (50 m.), available in 18 carat gold only, with a variety of dials.

Explorer 1016

Worn by Sir John Hunt in his conquest of Everest, the Explorer is a specially robust version of the fabled Oyster. Officially certified chronometer. Guaranteed waterproof to 330 feet (100 m.). Stainless steel case. Highly luminous dial.

Explorer II 1655

Known as the "Explorer II" this chronometer is of particular interest for speleologists. It is equipped with a 24 hour hand and the very strong winding crown is further protected by a special shouldering.

GMT-Master
1675

Created especially as an aviation timepiece, Pan American World Airways adopted this Oyster as the official chronometer for its pilots.

The GMT-Master indicates the precise time in two different time zones simultaneously. Perpetual rotor selfwinding. Automatic calendar. Guaranteed waterproof to 165 ft. (50 m.). Available in gold, steel and gold combination or stainless steel with matching bracelet.

Milgauss 1019

Thoroughly protected against magnetic fields of high intensity, it is a valuable precision instrument for use by scientists. Officially certified chronometer, waterproof, automatic Perpetual movement. In steel, with typical Rolex bracelet. Available with or without luminous dial.

Submariner 5513

As model 1680, the 5513 is equipped with the Triplock winding crown protected against damage by special shoulders. Fliplock extension bracelet permits watch to be worn outside of a light diving suit.

1680
Submariner-Date

Guaranteed waterproof to a depth of 660 feet (200 m.). Self-winding, officially certificated chronometer. Rotating bezel permits immediate measurement of elapsed time, an essential feature for the diver. Triplock winding crown. Shown here in 18 ct. yellow gold, with blue dial. Also available in steel.

Sea-Dweller 1665

A fine timepiece created for divers who work at depths up to 2000 feet (610 m.). This model has an escape valve for decompression which prevents the watch from exploding when surfacing. Yet another Rolex patent! Special Fliplock bracelet folding extension links.

6263

A stopwatch guaranteed to a depth of 165 feet (50 m.) in its Oyster case with screwed-down push buttons and Triplock winding crown.

The Rolex Cosmograph

The Rolex Cosmograph is available with engraved bezel or with black bezel in steel or in gold with various dials. Gold models are officially certified chronometers.

6265

Rolex Oyster for Ladies

Rolex was one of the first watch manufacturers to recognize that the day of the exclusively fragile women's watch was a thing of the past. Fancy models to suit a lady's fancy are still available and if anything they have more glamour than they used to. Yet, the active woman wants an attractive watch to serve her new professionnal status and responsibility.

With the Ladies Oyster, Rolex created just that kind of watch – incorporating all of the major achievements of the Oyster for men. The only difference – an important one – is that the Ladies Oyster is smaller, prettier, more feminine. Its classic, functional beauty is exactly right for any costume. The watch can be properly worn at work or at leisure and is correct even at a formal cocktail party.

For all its feminine glamour, the Ladies model is a pure-bred Rolex Oyster, with all the prestige, pleasure and technical perfection that the name implies. It is built exactly like its larger counterpart – with the seamless, indestructible Oyster case carved from a solid slab of steel or 18 ct. gold. And the Twinlock winding crown is guaranteed to keep out water under any conditions a woman might encounter. Inside the golden Oyster case is a hand-tuned, officially certified chronometer movement, with the famous rotor Perpetual selfwinding mechanism.

Rolex Oyster catalog
English version, 1973.
continued

At this point the eighteen-carat gold versions of the Cosmograph, models 6263 and 6265, carry the words *Superlative Chronometer Officially Certified* on the dial.

ROLEX

Your Rolex

Your Rolex booklet
English/North American
version, 1973.

This brochure gives important
dates in the development of the
Rolex Watch Company.
    It confirms the existence of a
Submariner watch in 1953.
    Note the reference to the 1954
launch date of the GMT-Master.

The 1971 launch date for the
Sea-Dweller, model 1665, may
refer to its promotion at a watch
fair. The French diving contractor,
Comex, had been using
prototypes of this watch since
1966 and it was sold to the
public from 1967.

1956

We launch the Rolex Oyster Perpetual "Day-Date",
the first waterproof, self-winding wrist chronometer
to indicate the date and the day of the week written
in full.

1960

A special Rolex Oyster fixed to the *outside* of the
bathyscaph "Trieste", dived to a depth of 35,798 feet,
where the pressure is nearly seven tons per square
inch. Upon surfacing, Jacques Piccard found that the
Rolex watch was functioning perfectly.

1971

Introduction of the Rolex Oyster Perpetual "Sea-
Dweller", guaranteed waterproof to a depth of
2000 feet—the first diver's watch incorporating a
helium valve.

1974

We have obtained by now more than 2 000 000 Timing
Certificates for wrist-chronometers—nearly half of
the entire production of Swiss chronometers to have
been officially certified to date by the Swiss Institutes
for Chronometer Tests.

18

Service Instructions

ROLEX LANDMARKS

1910

We obtain the first Official Timing Certificate for a
wrist-chronometer.

1914

At the world-famous Kew Observatory, we obtain the
first Class "A" Certificate ever awarded to a wrist-
watch.

1926

We develop and patent the world's first truly waterproof case, now universally known as the Rolex Oyster.

1931

We introduce the forerunner of all modern self-
winding wrist-watches, the famous Perpetual with the
self-winding rotor mechanism (Rolex patents).

16

1945

We introduce the "Datejust", the world's first
waterproof, self-winding calendar wrist-chronometer
automatically showing the date in a dial window.

1953

In the British Everest Expedition led by Sir John
Hunt, New Zealander Hillary and Sherpa Tensing
were the first to reach the summit of Everest. The
Expedition was equipped with Rolex Oyster Perpetual Chronometers which functioned with perfect
accuracy throughout despite extremes of temperature
and brutal shocks.

1953

    We launch the world's first wrist-watch especially
made for divers, the "Submariner".

1954

The first Rolex Oyster Perpetual Chronometer for
ladies is produced.

1954

    We introduce the Rolex Oyster Perpetual "GMT-
Master", a calendar chronometer designed specifically for pilots and giving a simultaneous reading of
the exact time in two different time zones.

17

Explorer owner's booklet
English/North American
version, early 1970s.

The Explorer, model 1016, and
the first Explorer II, model 1655,
are shown.
    The first Explorer II does not
have a luminous circle on the
seconds hand.

The Rolex Explorer

An officially certified
Rolex Oyster Perpetual
chronometer, with
selfwinding mecha-
nism, in a specially
treated stainless steel
case, guaranteed
waterproof to
a depth of 100 m.

Type I
Normal black face
with luminous num-
bers and plain bezel.

Type II
With supplementary
hand, plus specially
engraved "24 hour"
bezel and date calendar.

Ref. 1016/0   Ref. 1655/0

Explorer — the extra rugged Rolex Chronometer

The Explorer was developed at the time of the first successful expedition to Mount Everest, led by Sir John Hunt. By
then, Rolex Oyster Perpetual chronometers had equipped
no less than 14 expeditions to Everest and other peaks in
the Himalayas. By 1953 Rolex offered to explorers an even
more robust watch: the Explorer.
The Explorer is an officially certified chronometer, with a
selfwinding mechanism; protected by the famous Oyster
case, which is carved out of a solid block of specially
treated stainless steel. Screwed down into the Oyster case
is the exclusive Twinlock winding crown which works
exactly like a submarine hatch. The winder is the weak
point of ordinary watches, but the Twinlock makes the
Explorer perfectly waterproof and impermeable to dust.
The Explorer II which has only just been developed, will
be appreciated especially by speleologists who are prone
to lose all notion of time including day / night orientation
during their experiments in the depths of the earth.

1

The Explorer — a Rolex Red Seal Chronometer

The Rolex Red Seal is attached to each Rolex chronometer.
It signifies that the movement has successfully undergone
the severe tests for accuracy carried out by one of the
Official Swiss Institutes for Chronometer Tests. Carefully
scrutinised during 15 days and nights, in 5 positions and
at 3 specified temperatures, it fully justifies the jealously
guarded right to be called a *Swiss Chronometer*.
Even after this, the Explorer is rigorously tested in our own
laboratories before it is found worthy to bear the title,
Rolex Red Seal Chronometer. The proof that it has passed
these tests is found in the words "Superlative Chronometer
Officially Certified" which appear on the dial.

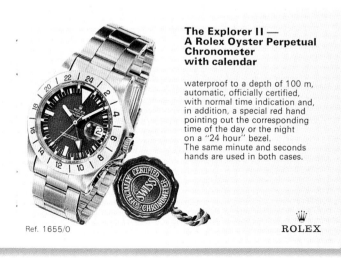

Ref. 1655/0

The Explorer II — A Rolex Oyster Perpetual Chronometer with calendar

waterproof to a depth of 100 m,
automatic, officially certified,
with normal time indication and,
in addition, a special red hand
pointing out the corresponding
time of the day or the night
on a "24 hour" bezel.
The same minute and seconds
hands are used in both cases.

ROLEX

2

These testimonials demonstrate that the Explorer can withstand the most extreme conditions

Expeditions to Mount Everest

From Sir John Hunt: " ... They performed splendidly and we have indeed come to look upon Rolex Oysters as an important part of high climbing equipment."

From Sir Edmund Hillary: " ... I wore this watch continously night and day. In the course of the expedition it experienced considerable extremes of temperature from the great heat of India to the cold temperatures at over 22,000 feet and seemed unaffected by the knocks it received on rock climbs or the continual jarring of long spells of step cutting in ice."

N. L. de Vall, Master, S/V Fri California

" ... I can honestly say that the Rolex Explorer is the ideal timepiece. It keeps accurate time, but that is not new or surprising. What is surprising is the phenomenal amount of hard wear it will stand up to and still keep accurate time."

4

The German Greenland Expedition 1970

From Mr. Peter Lechhart: " ... Since our radio direction finder had been rendered useless by disastrous atmospheric conditions, we were obliged to navigate with a sextant which requires an exact knowledge of the time. With joy and relief that we can assure you that our Rolex Chronometers never let us down."

New Zealand Antartic Expedition 1970

From Mr. R. Craig: " ... To sum everything up, I can say with complete honesty that this is the best watch in the world for any activity whatsoever. Conditions at Vanda are the most extreme in the world where man has, or is ever likely to set foot. Where else could you have a temperature range of 70° C and still have fine grit penetrating into every conceivable nook and cranny? Ruining cameras and sealed scientific instruments alike; I know of none on this earth, and yet this Rolex never faltered."

5

The Explorer I and the Explorer II

The Rolex Explorer I (ref. 1016) is a chronometer developed especially for daring men, to accompany them in hazardous conditions and has been proved in action time after time since 1953. Its performance has astonished people both for its accuracy and its robustness.

The Explorer II is a new watch. New in the sense that this model has been designed for the person who, in addition to the normal time indication, wants to know whether it is 10 o'clock a.m. or 10 p.m. (22 o'clock). This is the particular case of a speleologist.

A speleologist, in effect, is an explorer in the depth of the earth whose "daylight" is supplied by a pocket torch rather that by the sun. He soon loses all notion of time — morning, afternoon, day or night. The red hand of the Explorer II tells him in conjunction with the engraved bezel whether it is 2 o'clock a.m. or 2 o'clock p.m. and there is also a date on the dial.

6

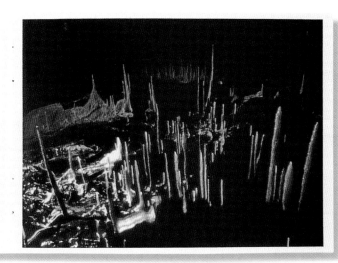

Submariner owner's booklet
English/North American
version, 1973.

This booklet accompanied both
the Submariner and Sea-Dweller
models.

The Sea-Dweller
is fitted with a valve (Rolex
patent) which allows helium
and other gases to escape,
enabling the watch to
withstand decompression.
This self-winding chrono-
meter with calendar has
been officially tested and is
guaranteed waterproof
to a depth of *2,000 feet*
(610 meters).

Ref. 1665/0

The Sea-Dweller Rolex Chronometer

Undersea exploration has developed to a remarkable extent
in the last few years. To enable him to work and to stay
under water for longer and longer periods, man has built
submarine houses and caissons in which the divers live in
an artificial atmosphere made of gas mixtures, usually with
a high percentage of helium. The very fine molecules of
such gases enter the watch but do not affect its operation.
However, during decompression, there is a risk that the
expanding gases will cause the watch to explode. To avoid
this by allowing the rapid escape of these gases, Rolex
have fitted a *patent* valve to their latest Oyster model, the
Sea-Dweller chronometer, which is guaranteed waterproof
to a depth of 2,000 feet (610 meters).

1

Ref. 5513/0   Ref. 1680/0

The Rolex Submariner

is guaranteed waterproof to a depth of 200 meters. The Triplock winding crown, a triple safety device (Rolex patent), is protected against damage by a shoulder. The rotating bezel is of vital importance when diving. The Submariner, with date, in 18 ct gold or in steel, is an *officially* certified chronometer.

Ref. 1680/8

The Rolex Submariner-Date

The Submariner is a handsome watch for business and social wear, yet remains, first and foremost, a diver's watch. The automatic movement with Perpetual rotor has obtained the *official* title of *Swiss Chronometer*. It is housed in a seamless Oyster case of 18 ct gold or surgical stainless steel. The Triplock crown, a triple safety device, is protected against all accidents by a seamless shoulder cut in the solid metal. The date changes promptly at midnight; and the bracelet, made from the same metal as the case, has a Fliplock fastening, with an extension, so that it can be worn over a light diving suit.

The dial and the rotating bezel of the gold Submariner are available in various colours, all very elegant. And while the bezel is useful for timing, say, a telephone conversation, its vital role is to indicate the decompression stages.

ROLEX

Ref. 1680/8

Rolex official dealer booklet
Europe/North American
version, 1976.

This booklet shows the sports
models.

174

SPÉCIALITÉS

4

6263/0   /8   Cosmograph
78350   7835/8

6265/0   /8   Cosmograph
78350   7835/8

6263/0   /8   Cosmograph
78350   7835/8

6265   /8   Cosmograph
78350   7835/8

5100/8/9   Quartz

10

1016/0   Explorer I
78360

1655/0   Explorer II
78360

1665/0   Sea-Dweller
93150

1675/0   /3   GMT-Master
78360   78363

1675/8   GMT-Master
6311/8

1019/0   Milgauss
78360

5513/0   Submariner
93150

1680/0   /8   Submariner Date
93150   9290/8

SPÉCIALITÉS
ROLEX OYSTER PERPETUAL

II

III

IV

V

VI

175

9

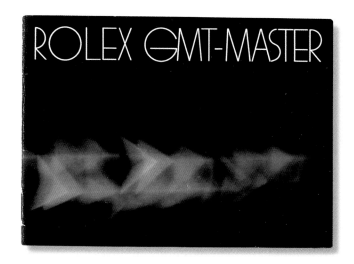

GMT-Master owner's booklet
English/North American
version, 1981.

The GMT-Master, model 16750,
previously known as the 1675
model, is featured below.
   To its right is the 16758 model.
On this watch the dial's luminous
markings were enclosed by gold
settings. During the 1980s, this
type of dial was introduced
across the entire sports model
range.

In 1986, this same type of dial
with white gold surrounds on the
dial's luminous markings was
fitted to the 16750 model.

THE GMT-MASTER CHRONOMETER

The GMT-Master Chronometer indicates simultaneously the
exact time in two chosen time-zones, according to the requirements of the moment. They may be GMT and New York
time, the time in Moscow and in Auckland, or any other two
places in the world. The GMT-Master Chronometer was
designed to meet the demands of the pilots and crews of two
world-famous airlines. It was an immediate success, and the
captains stated that the revolving bezel proved to be a very
valuable aid. The GMT-Master has also won the enthusiastic
approval of sea captains, members of the armed forces, and
business executives. The GMT-Master is an Officially Certified Chronometer, the movement is wound automatically by
the Perpetual rotor and the date can be rapidly adjusted. Its
Oyster case is guaranteed waterproof to a depth of 50 m
(165 feet). The revolving bezel and the 24 hour hand make
the GMT-Master an extremely versatile watch.   3

THE ROLEX GMT-MASTER

The GMT-Master is available
with different bezels and dials
as illustrated. The bezel was
designed to indicate the time
in a second time zone but
it can also be turned and
set to mark appointments
or used to measure elapsed
time such as the duration of
a long distance telephone
call. GMT-Master models in
18 ct gold are fitted with a
tough scratch resistant sapphire crystal.

16750   16758   16758

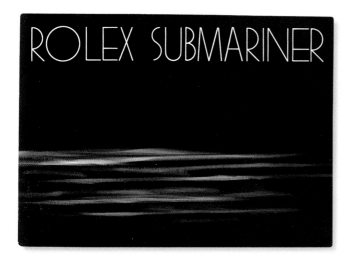

Submariner owner's booklet
English/North American
version, 1981.

Below, the Submariner, model
5513, which was fitted with an
acrylic lens.

Bottom, is the Submariner, model
16800, which at this stage was
fitted with a synthetic-sapphire
crystal but still had the old type of
dial. The dial's luminous markings
were not enclosed by white gold.

THE ROLEX SUBMARINER

is guaranteed waterproof to
a depth of 200 meters (660
feet). The Triplock winding
crown, a triple safety device
is protected against dam-
age by shoulders. The re-
volving bezel is of vital importance when diving.

5513

**THE ROLEX SEA-DWELLER
CHRONOMETERS**

ROLEX

Undersea exploration has developed to a remarkable extent.
To enable him to work and to stay under water for longer and
longer periods, man has built submarine houses and caissons
in which the divers live in an artificial atmosphere made of gas
mixtures, usually with a high percentage of helium. The very
fine molecules of such gases enter the watch but do not affect
its operation. However, during decompression, there is a risk
that the expanding gases will cause the watch to explode. To
avoid this by allowing for the rapid escape of these gases,
Rolex have fitted a patented valve to their Sea-Dweller
models. Rolex Sea-Dweller models are all Officially Certified
Chronometers. The self-winding Perpetual movements are
equipped with a calendar and fitted in special Oyster cases.
The two models available are illustrated overleaf.

7

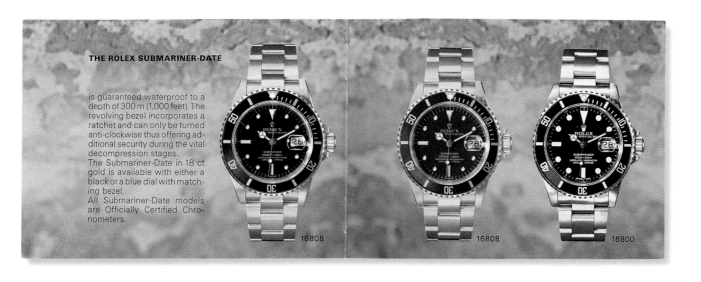

THE ROLEX SUBMARINER-DATE

is guaranteed waterproof to a
depth of 300 m (1,000 feet). The
revolving bezel incorporates a
ratchet and can only be turned
anti-clockwise thus offering additional security during the vital
decompression stages.
The Submariner-Date in 18 ct
gold is available with either a
black or a blue dial with matching bezel.
All Submariner-Date models
are Officially Certified Chronometers.

16808   16808   16800

Submariner owner's booklet
English/North American
version, 1981.
continued

Left, the Sea-Dweller, model
1665, with an all white-print dial
and an acrylic lens. It appears
alongside the Sea-Dweller, model
16660, which had a deeper
waterproof rating and was fitted
with a synthetic-sapphire crystal.

In late 1981, the 1665 model was
withdrawn from the Rolex
catalog. That it was on sale up to
this point suggests that the
watch was not a strong seller.

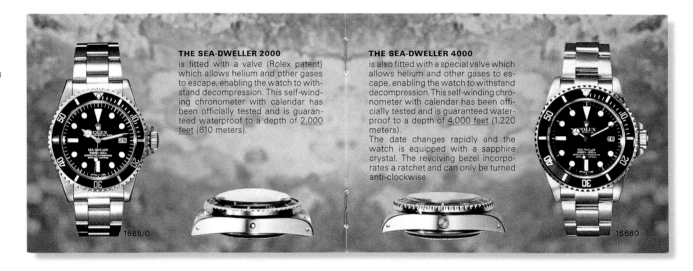

THE SEA-DWELLER 2000
is fitted with a valve (Rolex patent)
which allows helium and other gases
to escape, enabling the watch to withstand decompression. This self-winding chronometer with calendar has
been officially tested and is guaranteed waterproof to a depth of 2,000
feet (610 meters).

1665/0

THE SEA-DWELLER 4000
is also fitted with a special valve which
allows helium and other gases to escape, enabling the watch to withstand
decompression. This self-winding chronometer with calendar has been officially tested and is guaranteed waterproof to a depth of 4,000 feet (1,220
meters).
The date changes rapidly and the
watch is equipped with a sapphire
crystal. The revolving bezel incorporates a ratchet and can only be turned
anti-clockwise.

16660

THE PERPETUAL ROTOR **ROLEX**

invented and patented by Rolex in 1931, is the basis of all the
self-winding systems in mechanical wrist-watches. Constantly improved by Rolex over a period of more than 50
years, the Perpetual rotor means that the watch – even when
worn for as little as six hours a day – has a reserve operating
time of 48 hours.

THE ROLEX RED SEAL

and the words "Superlative Chronometer Officially Certified"
appearing on the dial prove that this Rolex has been awarded
the official title of Swiss Chronometer and has undergone
additional tests in our own laboratories. 11

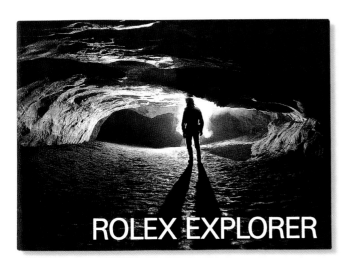

Explorer owner's booklet
English/North American version, 1986.

In mid-1985 the Explorer II, model 1655, was removed from the Rolex catalog and replaced by the 16550 model.

The new watch, shown below, was fitted with a synthetic-sapphire crystal, a new bezel design, and Mercedes hands.

The dial was available in either white or black. On the white dial watches the paint used was a pale ivory color, which faded to a distinctive cream which collectors have found very attractive.

These first models had white gold hands and white gold settings around the dial's luminous markings. By 1989 Rolex had changed the white dial paint to a non-fading bright white, and the bezel's numerals were represented with a new typeface. The settings around the luminous markings and the Mercedes hands were now painted black.

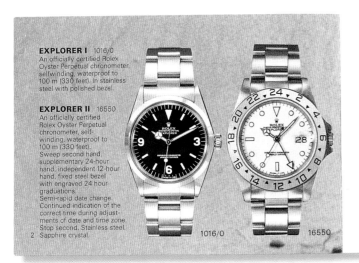

EXPLORER I 1016/0
An officially certified Rolex Oyster Perpetual chronometer, selfwinding, waterproof to 100 m (330 feet). In stainless steel with polished bezel.

EXPLORER II 16550
An officially certified Rolex Oyster Perpetual chronometer, selfwinding, waterproof to 100 m (330 feet). Sweep second hand, supplementary 24-hour hand, independent 12-hour hand, fixed steel bezel with engraved 24 hour graduations. Semi-rapid date change. Continued indication of the correct time during adjustments of date and time zone. Stop second. Stainless steel. Sapphire crystal.

1016/0   16550

"EXPLORER" MODELS
ROLEX

The Explorer I won its spurs during the first victorious expedition to Mount Everest in 1953. It is a Rolex Oyster Perpetual chronometer even more robust then the Oysters used successfully by fourteen previous expeditions to the Himalayas.

The Explorer II is a new Rolex Oyster Perpetual chronometer offering certain additional advantages, the date, 24-hour hand and 24 hour graduated bezel plus, a 12-hour hand that can be adjusted to allow for changes in time zones. Speleologists will find the 24 hour features useful during their explorations below the earth's surface when it is easy to lose all notion of time, especially day/night orientation. Long distance travellers will appreciate the fact that the 12-hour hand can show the time in one time zone while the 24-hour hand shows the time in another.

3

Chapter 17

Watches Sold at Auction
Christie's and Sotheby's

Watches Sold at Auction

The recent rapid growth in the number of people who collect wristwatches is due in part to the auction houses' decision to hold specialist wristwatch sales (which appeal more to men, who make up the majority of watch collectors), rather than, per previous practice, to include wristwatches in jewellery sales (which men may be less interested in attending). The success of these sales is due to two factors: a growing interest in wristwatches as jewellery, and an increase in the number of people who own more than one fine timepiece.

The following pages show images from Sotheby's and Christie's auction catalogs from the last four years. We have chosen to show the complete photographic plate which includes other non-sports models so that collectors will see the type of watches which frequently come up for auction.

Currently, along with general wristwatch sales, Christie's has a number of specialist Rolex-only sales and consequently we show slightly more plates from Christie's catalogues.

We have included the prices realised at auction for the sports models so that the change in watch values from 1997 to the present day may be seen. Prices at auction fluctuate more than the prices offered by dealers, and it can be seen that auctions are often a good place to find bargains.

We include a description of condition, described as high, medium or low range which relates to the price and condition guide on pages 208-11.

We are indebted to Richard Chadwick of Christie's UK, and Jonathan Darracott of Sotheby's UK, for locating and lending us these images.

Christie's London
26 June, 1997.

Top left, Explorer, model 1016.
Sold for £1995/$2992. Medium
range.

Center, Explorer 5504. Sold for
£1840/$2760. Low range. A later
dial has been fitted to this watch.

Top right, Explorer, model 1016.
Sold for £1840/$2760. Medium
range.

Bottom left, Explorer, model 6150.
Sold for £2300/$3450. Medium
range.

Bottom right, Explorer, model
6150. Sold for £2070/$3105.
Medium range.

184

© Christie's

Christie's London
26 June, 1997.

Center, Submariner, model 5508.
Sold for £552/$828. Low range.
 Bottom row, center, Turn-O-
Graph, model 6202. Sold for
£552/$828. Low range. This watch
has the correct early hands but has
significant dial damage.

Christie's London
2 October, 1997.

Left, Submariner, model 5513.
Sold for £1000/$1500. Medium
range. This watch is fitted with
the later style dial which has the
luminous markings on the dial
enclosed by white gold.

Right, Submariner, model 6538.
Sold for £900/$1350. Medium
range.

The watch at center is a diving
watch made for Panerai by Rolex in
the 1940s which was issued to the
Italian Navy in the 1940s.

© Christie's

Christie's London
27 November, 1997.

Top left, Explorer, model 5500.
Sold for £1300/$1950. Medium
range.

Top, second from left, GMT-
Master, model 1675. Sold for
£1500/$2250. Medium range.

Top right, Submariner, model
5513. Sold for £2800/$4200. This
is one of the military issue 5513
models.

© Christie's

Christie's London
27 November, 1997.

Explorer II, model 1655.
Sold for £4000/$6000. Medium
range. This watch was used by
Oliver Shepard in the Transglobe
Expedition of 1979-82. Shepard
instructed Rolex to fit a GMT-
Master dial to this watch. Such
custom-made pieces are rare.

Sotheby's Geneva
19 May, 1998.

Center row, left, Cosmograph, model 6239. Sold for £4571/$6856. Medium range.

Top left, Milgauss, model 6541. Sold for £14,435/$21,652. Medium range.

Top right, Submariner, model 1680. Sold for £1443/$2165. Medium range.

The watch at center is a Cosmograph, model 6238, which was the precursor of the 6239 model.

Sotheby's London
11 June, 1998.

Bottom right, Milgauss, model
1019. Sold for £4600/$6900.
Medium range.

Christie's London
24 June, 1998.

Top left, Submariner, model 5512. Sold for £1150/$1725. Medium range.

Top right, Submariner, model 5513. Sold for £2300/$3450. Medium range. This is one of the military issue 5513 models. The original hands have been replaced by Mercedes hands.

Center row, left, Submariner, model 5513. No sale.

Center row, right, Turn-O-Graph, model 6202. Sold for £1265/$1897. Medium range.

Bottom left, Submariner, model 5513. No sale. This is one of the military issue 5513 models. The original hands have been replaced by Mercedes hands.

Bottom right, Submariner, model 6536/1. Sold for £977/$1465. Medium range.

Christie's London
24 June, 1998.

Top, second from left, Explorer II, model 16550. Sold for £2645/$3967. Medium range.
    Bottom, second from left, GMT-Master, model 1675. Sold for £2070/$3105. Medium range.
    Bottom right, GMT-Master, model 1675. Sold for £1265/$1897. Medium range.

Christie's London
24 June, 1998.

Top left, Submariner, model 6538. Sold for £1725/$2587. Low range This watch has a restored dial.

Top, second from left, Submariner, model 5513. No sale. Medium range.

Top right, Submariner, model 5513. No sale. Medium range.

Center row, second from left, Explorer, model 6350. No sale. Medium range.

Center row, second from left, Submariner, model 5513. Sold for £3680/$5520. Medium range. This is a Comex issue 5513 with a gas escape valve fitted.

Center row, right, Explorer, model 5500. Sold for £1380/$2070. Medium range.

Bottom left, Explorer, model 1016. Sold for £2185/$3277. Medium range.

Bottom, second from left, Explorer, model 5504. No sale. Medium range.

Bottom right, Explorer, model 5500. Sold for £977/$1465. Medium range.

Sotheby's London
26 February, 1998.

Bottom right, Cosmograph, model
6239. Sold for £8625/$12,937.
Medium range.
   The watch at top left is the later
style Cosmograph which was
introduced in the early 1990s.

Sotheby's London
26 February, 1998.

Top right, Explorer II, model 1655.
No sale. Medium range.

Sotheby's New York
27 October, 1998.

Top left, Cosmograph, model 6263.
Sold for £5366/$8050. Medium
range.
    Top, second from left,
Cosmograph, model 6262. Sold for
£4216/$6325. Medium range.

Christie's London
25 November, 1998.

Top right, GMT-Master, model
6542. Sold for £1600/$2400.
Medium range.

Christie's London
24 February, 1999.

Bottom left, Explorer, model 1016.
Sold for £2000/$3000. Medium
range.
   Bottom, second from left,
Submariner, model 5512. Sold for
£1100/$1650. Medium range.

198

Sotheby's London
23 March, 1999.

Top, second from left, Submariner, model 6536. No sale. Medium range.

Center row, left, Turn-O-Graph, model 6202. Sold for £1265/$1897. Medium range.

Center row, second from left, Submariner, model 5508. Sold for £1495/$2242. Medium range.

Center row, right, Submariner, model 5513. Sold for £2300/$3450.

Medium range. This is one of the military issue 5513 models.

Bottom left, Explorer, model 5504. Sold for £1380/$2070. low range This watch has a later dial which does not have *Explorer* on it - this is a replacement dial issued by Rolex and was never fitted to any model.

Bottom, second from left, Submariner, model 5513. Sold for £4600/$6900. Medium range. This is a Comex issue 5513 with a gas escape valve fitted.

Sotheby's Geneva
18 May, 1999.

Top left, GMT-Master, model 1675.
Sold as part of a lot. Medium
range.
   Top right, GMT-Master, model
6542. Sold for £1684/$2526.
Medium range.

200

Sotheby's New York
15 June, 1999.

Top left, Sea-Dweller, model 1665.
Sold for £6516/$9775. Medium
range. This is a Comex issue Sea-
Dweller.

Top right, Cosmograph, model
6240. Sold for £6900/$10,350.
Medium range.

Bottom second from left,
Explorer, model 1016. Sold for
£4600/$6900. Medium range. This
Explorer is signed *Space Dweller*.

Christie's London
16 June, 1999.

Top left, GMT-Master, model 1675.
Sold for £1400/$2100. Medium
range.
  Center row, second from left,
GMT-Master, model 1675. Sold for
£1200/$1800. Medium range.
  Bottom left, Submariner, model
1680. Sold for £1408/$2112.
Medium range.

Christie's London
16 June, 1999.

Top, Explorer, model 1016. Sold for
£1700/$2550. Medium range.

203

© Christie's

Christie's London
16 June, 1999.

Top left, Submariner, model 6536.
Sold for £1050/$1575. Medium
range.

Top right, Submariner, model
5508. Sold for £1600/$2400.
Medium range.

Center row, Explorer, model
5500. Sold for £2000/$3000.
Medium range.

Bottom left, Submariner, model
1680. Sold for £1400/$2100.
Medium range.

Bottom, second from left,
Submariner, model 1680. Sold for
£1800/$2700. Medium range.

Bottom right, Sea-Dweller,
model 1655. Sold for
£1700/$2550. Medium range.

Sotherby's London
19 December, 2000.

Top, Explorer, model 1016. Sold for
£2400/$3600. Medium range.
   Center row, left, Submariner,
model 5513. No sale. Medium
range.
   Bottom left, Submariner, model
1680. No sale. Medium range.

Those wishing to order auction house catalogs
may find the following addresses usefull:

Sotheby's

England
34-35 New Bond Street
London
W1A 2AA
Iel: 020 7293 5000

France
76 Rue de Fauborg St.
Honoré
75008
Paris
Tel: 33 1 53 05 53 05

Italy
Palazzo Broggi
Via Broggi 19
20129 Milan
Tel: 39 02 29 5001

Germany
Mendelssohnstrasse 66
D-60325 Frankfurt-am-main
Tel: 49 69 74 07 87

USA
1334 York Avenue
New York, New York 10021
Tel: 212 606 7000

Christie's

England
85 Old Brompton Road
London
Sw7 3LD
Tel: 020 7581 7611

France
9 Avenue Matignon
75008
Paris
Tel: 33 1 40 76 85 85

Italy
Piazza Santa Maria delle Grazie 1
20123 Milano
Tel: 39 02 46 70 141

Germany
Giesebrechtstraße 10
10629
Berlin
Tel: 49 30 88 56 95 0

USA
20 Rockefeller Plaza
New York, New York 10020
Tel: 212 636 2000

Chapter 18

Price Guide

Price guide, and a guide on grading watches according to condition.

Condition Guide

Below are some suggestions which the potential buyer can bear in mind when examining a watch under consideration for purchase. A loupe (magnifying eyepiece) is an indispensable tool for examining a watch. The most important condition factors are:

Dial
• It is wise to be familiar with which type of dial that should be on the watch in question.
• The dial should have the original luminous markings, preferably be unrestored, and have no damage.
• Check that the brightness of the luminous markings on the hands matches that of the luminous markings on the dial. If they are both original, the color should be the same.
• A well-restored dial is more valuable than a badly damaged dial. Damaged dials are best avoided.

Bezel
• Be familiar with the type of bezel which should be on the watch.
• Beware of damaged bezels - they are impossible to replace on older models.

Watchcases (steel cases only):
• Watchcases must be original and in good order. Small scratches can be polished clean, and a Rolex Service Center will do the best job with the correct tools.
• Heavily pitted and dented cases are best avoided as they cannot be repaired. Rolex can supply new cases for later models, but older cases have been discontinued.
• Beware of highly-polished cases. They may look attractive, but they are worth less because vigorous polishing wears away the watchcase's original curves and sharpness.

Hands
• These should be original and unrestored, with the original luminous paint intact. On an older watch it is common to find some of the luminous paint missing from the hands. Note that it is nearly impossible to restore the hands so they match the color the dial's other luminous markings.

Winder
• This should be fully functioning, and be

Explorer
Manufactured in 1986.

Model Number: 1016
Case number: 9664504

1 Cardboard outer box.
2 Inner box.
3 Brochure.
4 case back sticker, this has been removed and stuck to the front of the Chronometer certification.
5 Chronometer certification.
6 Translation of the Chronometer certification.
7 Calendar.

the correct winder for the model in question. Winders for some older models are impossible to find.

Lens
• This lens should be the correct one for the model in question, and be free of cracks or chips. Small scratches on the surface of the lens can be removed with cotton wool and toothpaste.

Bracelet
• Links should be free of excessive stretch, and should be original. Older, already-stretched bracelets are best avoided. If an old bracelet is small, it will be difficult to extend, as replacement links may not available.
• Flush-fit end pieces should be the correct size for the model.

Leather straps
• These should be correctly signed by Rolex, with original clasps. Watch for fake leather straps and clasps, which are usually poorly made. Fake clasps are made from softer metals than steel.

Box and papers
• "Box and papers" means at minimum that the watch should have its original guarantee, and both the cardboard outer box and wooden inner box.

Movements
• Movements should be undamaged. Screw back, waterproof watches provide greater protection for movements, so the watches in this book are less likely to suffer movement damage. Damage typically comes from water, when people forget to screw the winder back on.
• Non-working watches are best avoided, as repairs could be impossible or very costly. The old maxim 'if it ticks it can be fixed' is usually not true. A reputable dealer will always sell a watch in working order.

Sea-Dweller
Manufactured in 1977.

Model Number: 1665
Case number: 5067418

1 Cardboard outer box.
2 Inner box.
3 Anchor - marked with the depth rating of the watch.
4 Chronometer certification.
5 Translation of the Chronometer certification.

Price Guide

Three price categories are described below. Each has been based on the watch's condition.

High range

This is for mint condition pieces with all original features intact and no damage whatsoever e.g. original papers, boxes, bracelet, lens, bezel and all luminous paint on the dial and hands matching; the watches shown on pages 208 and 209 would fall into this category.

Medium range

Most dealers would also describe this category as mint. This type of watch would be in good condition, but might lack papers or some bracelet links. Overall, it might have show some wear, but the dial would probably be in a good state. These watches will suit many if not most collectors, as perfect, high-priced pieces are less often on the market.

Low range

Despite attractive prices, these watches are not best suited to collectors. They might have significant damage, and are better suited to dealers, who might have access to the necessary parts. A watch in this category could have a refinished dial, a highly polished case, new hands, and an incorrect bezel and lens. Most reputable dealers offer watches in a condition that commands a medium price range. Very few deal in high price range watches. It is possible, though rare, to find a correctly restored or original condition watch in the medium price range. Apparent bargains must be thoroughly checked, because there is almost always something unattractive behind an attractive price. Damaged pieces for sale at the right price can be restored to excellent condition by a Rolex service center, but the cost can be substantial.

The prices in the following price guide are correct as of May 2001, page numbers are given for watches that appear in this book. If a watch is not shown in this book the page number is given as N/S.

It should be noted that this price guide is the opinion of the authors, and neither the authors or publisher may be held responsible in any way for loss or gain incurred by using this guide.

Condition check list for a high range watch with a rotating bezel
1 Correct case polishing with a polished bevel separating the polished sides from the satin finished top.
2 Correct size flush-fit endpiece for this model
3 Original and correct bezel for this model, unscratched or retouched.
4 Original dial with all luminous markings intact.
5 Original hands with all luminous paint intact. The luminous paint should match the color of the luminous paint on the dial.
6 Correct lens for this model. Free from chips and cracks.

Condition check list for a high range watch with a fixed bezel
1 Correct case polishing with a crisp edge between the polished sides and satin finished top.
2 Correct size flush-fit endpiece for this model.
3 Original dial with all luminous markings intact.
4 Original hands with all of the luminous paint intact. The luminous paint should match the catalog of the luminous paint on the dial.
5 Correct lens for this model. Free from chips and cracks.

Submariner	High £	Medium £	Low £	High $	Medium $	Low $	Page
Deep Sea Special/Piccard	£75,000	£50,000	N/A	$112,500	$75,000	N/A	26
6205	£5000	£3500	£1200	$7500	$5250	$1800	26
6200 Explorer dial	£8500	£7000	£2500	$12,750	$10,500	$3750	27
6200 Standard dial	£7500	£6000	£2000	$11,250	$9000	$3000	34
6204	£5000	£3500	£1200	$7500	$5250	$1800	27
6536	£5000	£3500	£1200	$7500	$5250	$1800	30
6536/1 Chronometer version	£6000	£4500	£1500	$9000	$6750	$2250	31
6538	£5000	£3500	£1000	$7500	$5250	$1500	32
6538A	£6000	£4500	£1500	$9000	$6750	$2250	32
5510	£7500	£5000	£2000	$11,250	$7500	$3000	34
5508	£6500	£4500	£1500	$9750	$6750	$2250	35
5512 Explorer dial	£9500	£7500	£3000	$14,250	$11,250	$4500	N/S
5512 Standard dial	£3500	£2500	£900	$5250	$3750	$1350	36
5513 Explorer dial	£7500	£6000	£2000	$11,250	$9000	$3000	37
5513 Standard dial	£2500	£1750	£900	$3750	$2625	$1350	37
1680 Red writing	£4000	£3000	£1000	$6000	$4500	$1500	42
1680 White writing	£2500	£1750	£900	$3750	$2625	$1350	44
5513 Military	£8500	£6500	£1500	$12,750	$9750	$2250	42
5517 Military	£8500	£6500	£1500	$12,750	$9750	$2250	43
16800 Synthetic sapphire lens	£2500	£1800	£900	$3750	$2700	$1350	45

211

Sea-Dweller	High £	Medium £	Low £	High $	Medium $	Low $	Page
5513 Comex	£8500	£7000	£2500	$12,750	$10,500	$3750	52
5514 Comex	£10000	£8000	£2500	$15,000	$12,000	$3750	53
1665 Red writing	£6500	£5000	£1500	$9750	$7500	$2250	53
1665 White writing	£5500	£3500	£1000	$8250	$5250	$1500	55
16660 Sea-Dweller	£3000	£2500	£900	$4500	$3750	$1350	55
16660 Comex Sea-Dweller	£8500	£7000	£2000	$12,750	$10,500	$3000	N/S

GMT-Master	High £	Medium £	Low £	High $	Medium $	Low $	Page
6542 Acrylic bezel	£4500	£3500	£1250	$6750	$5250	$1875	60
6542 Metal bezel	£3000	£2250	£950	$4500	$3375	$1425	61
6542 Pan American white dial	£8500	£7500	£1000	$12,750	$11,250	$1500	62
1675	£3000	£2250	£950	$4500	$3375	$1425	63
16750	£3000	£2250	£950	$4500	$3375	$1425	67

continued overleaf

Explorer	High £	Medium £	Low £	High $	Medium $	Low $	Page
6098	£3500	£2500	£950	$5250	$3750	$1425	74
6098 Honeycomb dial	£4000	£3000	£1000	$6000	$4500	$1500	N/S
6298	£4500	£3500	£1000	$6000	$5250	$1500	74
6298 Honeycomb dial	£5000	£4000	£1500	$7500	$6000	$2250	N/S
6150	£3500	£2500	£950	$5250	$3750	$1425	N/S
6150 Honeycomb dial	£4000	£3000	£1500	$6000	$4500	$2250	76
6350	£4500	£3500	£1000	$6000	$4500	$1500	75
6350 Honeycomb dial	£5000	£4000	£1500	$7500	$6000	$2250	N/S
6610	£4500	£3500	£1000	$6000	$5250	$1500	76
6610 Honeycomb dial	£5000	£4000	£1500	$7500	$6000	$2250	77
1016 Hacking movement	£4000	£3250	£2750	$6000	$4875	$4125	79
1016 Non-hacking movement	£3500	£2750	£950	$5250	$4125	$1425	N/S
1016 Space-Dweller	£7500	£6000	£1500	$11,250	$9000	$2250	82
5500 Gloss dial	£3500	£2500	£950	$5250	$3750	$1425	78
5500 Standard dial	£2500	£2000	£950	$3750	$3000	$1425	80
5504 Gloss dial	£3000	£2500	£950	$4500	$3750	$1425	N/S
5504 Standard dial	£2500	£1650	£800	$3750	$2475	$1200	79
1655 Explorer II	£5500	£4000	£1250	$8250	$6000	$1875	84
16550 Explorer II with cream dial	£4500	£3750	£1250	$6750	$5625	$1875	86

Milgauss	High £	Medium £	Low £	High $	Medium $	Low $	Page
6541 *Milgauss* on lower half of dial	£30,000	£20,000	£4000	$45,000	$30,000	$6000	92
6541 Submariner bezel	£15,000	£10,000	£2500	$22,500	$15,000	$3750	92
6541 Submariner bezel & Honeycomb dial	£16,500	£12,000	£3000	$24,750	$18,000	$4500	N/S
6541 Milgauss bezel	£20,000	£15,000	£3000	$30,000	$22,5000	$4500	93
6541 Milgauss bezel & Honeycomb dial	£21,500	£16,000	£3500	$32,250	$24,000	$5250	94
1019 1960s	£10,000	£7500	£2000	$15000	$11,250	$3000	95
1019 1970s-80s	£8500	£6500	£1800	$12,750	$9750	$2700	96

Turn-O-Graph	High £	Medium £	Low £	High $	Medium $	Low $	Page
6202	£8000	£4000	£1000	$12,000	$6000	$1500	102
6202 Honeycomb dial	£9000	£4500	£1500	$13,500	$6750	$2250	102

212

Cosmograph	High £	Medium £	Low £	High $	Medium $	Low $	Page
6239 Standard dial	£8500	£7000	£2500	$12,750	$10,500	$3750	112
6239 Standard all silver dial	£9000	£7500	£3000	$13,500	$11,250	$4500	114
6239 Exotic dial black with white recording dials	£15,500	£13,000	£5000	$23,250	$19,500	$7500	N/S
6239 Exotic dial white with black recording dials	£13,500	£11,000	£3500	$20,250	$16500	$5250	112
6241 Standard dial	£8500	£7000	£2500	$12,750	$10,500	$3750	N/S
6241 Exotic dial black with white recording dials	£15,500	£13,000	£5000	$23,250	$19,500	$7500	113
6241 Exotic dial white with black recording dials	£13,500	£11,000	£3500	$20,250	$16500	$5250	N/S
6240 Standard dial	£8500	£7000	£2500	$12,750	$10,500	$3750	N/S
6240 Exotic dial black with white recording dials	£15,500	£13,000	£5000	$23,250	$19,500	$7500	N/S
6240 Exotic dial white with black recording dials	£13,500	£11,000	£3500	$20,250	$16500	$5250	116
6262 Standard dial	£8500	£7000	£2500	$12,750	$10,500	$3750	N/S
6262 Exotic dial black with white recording dials	£15,500	£13,000	£5000	$23,250	$19,500	$7500	115
6262 Exotic dial white with black recording dials	£13,500	£11,000	£3500	$20,250	$16500	$5250	N/S
6264 Standard dial	£8500	£7000	£2500	$12,750	$10,500	$3750	N/S
6264 Exotic dial black with white recording dials	£15,500	£13,000	£5000	$23,250	$19,500	$7500	115
6264 Exotic dial white with black recording dials	£13,500	£11,000	£3500	$20,250	$16500	$5250	N/S
6263 Standard dial	£8500	£7000	£2500	$12,750	$10,500	$3750	117
6263 Exotic dial black with white recording dials	£15,500	£13,000	£5000	$23,250	$19,500	$7500	N/S
6263 Exotic dial white with black recording dials	£13,500	£11,000	£3500	$20,250	$16500	$5250	N/S
6265 Standard dial	£8500	£7000	£2500	$12,750	$10,500	$3750	N/S
6265 Exotic dial black with white recording dials	£15,500	£13,000	£5000	$23,250	$19,500	$7500	N/S
6265 Exotic dial white with black recording dials	£13,500	£11,000	£3500	$20,250	$16500	$5250	116

Index

Bold type indicates an Illustration

Air-King: Explorer model 5500. see Explorer model 5500
Alitalia, 58
alloys: anti-magnetic, 90, 91, 95
Apollo astronauts, 59
Armstrong, Neil, 59
arrowhead markers, 70, 72, 91, 92, 93
Auguste Piccard. See Piccard
automatic rotary winding mechanism, 10

balance mechanism: Superbalance, 11
Bangkok, 132
Basel Spring Watch Fair, 11, 20, 21, 27
Bathyscaphe: FNRS-2, 11, 25
Bathyscaphe: Trieste, 12, 25
baton markers, 91, 95, 96, 97
Bavaria, 8
bezel, 10, 11, 12, 17, 24, 58, 73, **84**, **86**, 91, 100, 101, 102, 108, 109, 113, 132, 134, 208; Cosmograph bezel, **112**, **113**; bezel insert, 58, 59, **60**, 61, **61**, 62, **66**, 109, 113; Milgaus bezel, 90, 91, 93, **93**, 134; bezel ring, 58, 62, **63**,108; Sea-Dweller bezel, 49; Submariner bezel, 20, 22-25, **26**, 31, **31**, **35**, 43, **43**, 58, 90, 92, **92**, 134; Turn-O-Graph bezel, 15, **102**, **105**, 108, 134
Bienne: Switzerland, 9
Bluebird, 10
Bond, James, 23
Bourdillon, Tom, 71
box, 122-125, **123-125**, 209
bracelet, 24, 32, 118-121, 209
bracelet: Jubilee, 59, 66, 118, **119**, **120**, **121**
bracelet: Oyster, 59, 66, 83, 118, **119**, **120**, **121**
Brevet, 20, 26
Brevette, 20
bubbleback, 11, 14, 70, 74

Canadian. see Navy
Campbell, Sir Malcom, 10
Chaux de Fonds, La, 8, 9
Christie's, 25
Cho Oyu, 70
Compressed air. See diving
Cosmograph, 17, 108-117, 129, 134
    model 6238, 108
    model 6239, 12, **17**, 108, **108**, 112, **112**, 113, **113**, 114, **114**, 129
    model 6240, 13, 109, **109**, 116, **116**, 129
    model 6241, 108, **108**, 113, **113**, 114, **114**, 129
    model 6262, 108, 115, **115**, 117, 129
    model 6263, 109, 110, 117, **117**, 129
    model 6264, 109, 115, **115**, 129
    model 6265, 109, 110, 116, **116**, 129
Chronograph, 59, 101, 106, 108, 109
Chronograph: waterproof, 11, 101
Chronometer, 9, 11, 13, 22, 31, 36, 102, 103, 109, 132
Chronometer: First Class Certificate, 9, **127**

Chronometer: Officially Certified, 23, 33, 36, 74, 76, 102, 103, 105, 126
Chronometer: Superlative; Officially Certified, 23, 38, 79, 109, 126
Comex, 16, 48, 49, 52; Comex identification number, 49, 52; Comex logo, 49, 50, 52, 53, 54,
    model 5513, 128
    Sea-Dweller 1665. See Sea-Dweller
    Sea-Dweller 16660. See Sea-Dweller
    Submariner 5514. See Sea-Dweller
Cousteau, Jacques, 15
crown guards, 12, 20, 23, **23**, 36, 38, 59, 63, 65
crown logo, 9, **9**, 63, 85
Cyclops lens, 58, 61

Daily Mail, 10
Daytona, 131
Davis, Alfred, 8, 9
decompression, 24, 48, 49
decompression illness, 20, 24
Deep Sea Special, 11, 25, **25**, 26, **26**, 128
depth rating, 14, 15, 20-22, 27, 28, 29, 31, 33, 35, 39, 45, 50, 53-55, 58, 60, **71**, 77, 100, 101, 102, 104
dial: aluminum, 91, 95; blank, 22; exotic, 110, 112, **112**, 113, 115, 116, 117, 133; Explorer-type dial, 23, 26, 34, 37, 43, 71; honeycomb finish dial, 71, 76, 78, 91, 94, 100, 101, 105; Paul Newman dial. See Newman. Quarter Arabic dial, 71, 74; Sea-Dweller dial, 54; Submariner dial, 34, 90
dial marking: Cosmograph, 108, 110, 117, Daytona, 108, 110, 117; Explorer, 75, 76, 77; Milgauss, 90, 92; Oyster, 109, 116; Oyster Perpetual Date, 50, 54, 59; O T Swiss T O, 116; Precision, 71, 72, 74, 78, 83; Rolex, 81, 82, 85; Submariner, 20, 21, 24; Super Precision, 71, 72, 78, 79, 81, 83; Swiss T<25, 38, 39, 65, 79, 81; T (tritium), 24, 42; T Swiss T, 117; Turn-O-Graph, 100, 102
diving: compressed air, 48
diving: saturation, 13, 16, 48, 50, 52, 54
diving: scuba, 15, 20, 21

Everest, 11, 15, 70, 74
Explorer, 11, 14, 15, 16, 22, 70-87, 91, 100, 118, 129, 132, 134, **208**
    model 1016, 71, **71**, 72, **72**, 79, **79**, 81, **81**, 82, **82**, 83, **83**, 86, **86**, 87, **87**, 122, 129, 134, **208**
    model 1016: Space-Dweller, 72, **72**, 82, **82**, 129
    model 5500: Air-King, 72, **72**, 78, **78**, 80, **80**, 81, **81**, 83, **83**, 129
    model 5504, 71, 79, **79**, 129
    model 5701: Explorer date, 72
    model 6098: prototype, 11, 70, **70**, 71, 74, **74**, 129
    model 6150, **14**, 15, 71, 75, **75**, 129
    model 6298, 21, 71, **71**, 72, 74, **74**, 129
    model 6299, 72

model 6350, 71, 75, **75**, 76, **76**, 91, 129
model 6610, 71, **71**, 76, **76**, 77, **77**, 78, **78**, 129
model 14270, 72
Explorer II: model 1655, 13, 17, **17**, 72, 73, **73**, 84, **84**, 85, **85**, 86, 129, **129**,134
Explorer II: model 16550, 73, **73**, 86, **86**, 129

fake, 132, 133, 134
First World War watch, 14

gas: escape valve; helium, 13, 16, 49, **51**, 52, **52**, 55
gauss, 90
Geneva, 9, 59
Gleitze, Mercedes, 10
GMT-Master, 11, 12, 14, 16, 57-67, 101, 118, 122, 128, 134
    model 1675, 59, **59**, 63, **63**, 64, **64**, 65, **65**, 66, **66**, 67, 126, 128
    model 6542, 12, **16**, 58, **58**, 60, **60**, 61, **61**, 62, **62**, 65, **65**, 126, 128
    model 16750, 59, 67, **67**, 128
GMT-Master II, model 16750, 59
gold, 24, 58, 59, 100, 109, 110
Gruen, 133

hacking, 71
hands: angled profile, 91, 95
hands: Mercedes, 22, 30, 58, 71, 73, 75, 77, 86, 90, 91, 94, 100, 103, 104
hands: leaf-shaped, 70, 90, 92
hands: hour; secondary, 58, 63, **63**
hands: pencil-shaped, 20, 71, 76, 100
hands: seconds; lightning bolt, 90, 91, 93; painted, 31, 33, 31, 77, 103, 109, 114
hands: Submariner, 22
Hatton Garden, 8
Hillary, Sir Edmund, 15, 70
Hong Kong, 132
Horology: Society of, 9
Hunt, Sir John, 11, 70, 74

Italian. see Navy

Japan, 72, 82
jewels, 23, 128-129

King George V, 10

Lambert, Raymond, 70
lens: acrylic crystal, 23, 25, 48, 49
lens: synthetic sapphire crystal, 25, 45, 50, 55, 73, 86
Lindbergh, Charles, 14
London, 8
Longines, 14
Luminova, 65

magnetic fields, 12, 90
Marines, British Royal, 24, 43
Mercedes. See hands

214

Mercury astronauts, 59, 72
Mercury Space Programme, 72, 82
Milgauss, 16, 17, 22, 90-97, 101,129
    model 1019, 91, **91**, 94, 95, **95**, 96, **96**,
    97, **97**, 129
    model 6541, 12, **16**, 22, 90, **90**, 91, 92, **92**,
    93, **93**, 94, **94**, 95, 129
mille, 90
minute marks, 22, 24, 37, 40, 43, 44, 66, 79,
80, 100
Montres Rolex, 9

NASA, 59, 72, 72, 82, 109
Navy, British Royal, 24, 28, 32, 42
Navy, Italian, 14
Navy, Royal Canadian, 24, 32
Newman, Paul, 110, **110**, 112, 133
Neuchatel, La Chaux de Fonds, 8
Norgay, Jamling, 70
Norgay, Tenzing; Sherpa, 15, 70

oersted, 90
Omega, 59, 109
Omega Speedmaster: Moonwatch, 59
Oyster, 10, 11, 15, 71
Oyster: patent, 10
Oyster Perpetual, 10, 11
Oyster Perpetual Datejust, 11
Oyster Perpetual Day-Date, 12
Oyster Zerograph, models 3642, 3346, 10

Pan American Airlines; Pan Am, 16, 58, 62
Panerai Company, 14
paperwork, 50, 126-127, **126-127**, 132, 209
Piccard, Auguste: Professor, 11, 25
Piccard, Jacques, 25
Piccard, model 7205/0, 25, **25**, 26, 128
Precision, models 3003, 3004, 10
pressure relief valve, 49, **51**, 52, 53
Prince, 10, 14, 133

Rolex Watch Company, 8, 14
Rolex Watch Company Ltd, 9
Royal Geographical Society, 70
Royal Navy: British. See Navy.

saturation. See diving
scuba. See diving
screw down crown, 10, **10**, 25
screw down pusher, 109, 116
Sea-Dweller, 47-55, 118, 119, 126, 128, **209**
    model 1665, 13, **16**, 49, **49**, 50, 51, 53,
    53, 54, **54**, 55, **55**, 126, 128, 209
    model 1665: Comex, 50, **50**, 54, **54**; case
    back, **51**, **54**
    model 5514: Submariner; Comex, 49, **49**, 53,
    53, 128, 134
    model 16660: Comex, 50, **50**, 55, **55**, 128,
    134; case back, **51**, **55**
Sea-Dweller 2000: Submariner, 16, 50, 55
service numbers: military watches, 24, 32, 42
Silver Jubillee: King George V, 10

Smiths watches, 70
Space-Dweller: Explorer model 1016. See
Explorer model 1016.
speleologist, 17, 73
strap: non-reflective watch strap, 24, 32, 42
strap bars, 24, 32, 42
steel, 11
steel & gold, 11
Submariner, 11, 14-17, 19-45, 58, 90, 100, 101,
118, 119, 122, 128, 133, 131, 133, 134, 135
    model 1608, 13, 23, **23**, 25, 42, **42**, 44, **44**,
    45, 128, 134
    model 5508, 23, 35, **35**, 39, **39**, 40, **40**, 41,
    41, 128
    model 5510, 22, 34, **34**, 40, **40**, 128
    model 5512, 12, 20, 23, **23**, 36, **36**, 38, **38**,
    41, **41**, 43, 128, 131
    model 5513, 23, 24, 25, 37, **37**, 38, **38**, 39,
    39, 42, **42**, 43, **43**, 44, **44**, 48, 49, **51**, 52,
    52, 128, 131; Comex, 48, 49; non gas-
    escape valve model, 51; case back, **51**, **52**
    model 5514: Submariner; Comex. See Sea-
    Dweller
    model 5517, 24, **24**, 43, **43**, 128
    model 6200, **15**, 20-23, **20**, 26, **26**, 30, **30**,
    32,34, **34**, 126, 128
    model 6204, 20-22, **20**, 21, 24, 27, **27**, 28,
    28, 58, 126, 128
    model 6205, 20-22, 26, 27, 28, **28**, 29, **29**,
    30, 128
    model 6536, 22, 23, 30, **30**, 31, **31**, 41,
    128
    model 6536/1, 22, **22**, 31, **31**, 35, 128
    model 6538, 22, **22**, 23, **23**, 24, 29, **29**, 32,
    32, 33, **33**, 41, 128
    model 6538A, 22, 32, **32**, 128
    model 16800, 25, 45, **45**, 128
Submariner Sea-dweller 2000. See Sea-Dweller
2000
superbalance, 11
Swiss movement: imports, 8
Switzerland, 9, 11

tachymetric timing ring, 12, 108
technical office: La Chaux de Fonds, 8
Tenzing Norgay. See Norgay.
time zones, 58
Tritium, 24, 42, 65
Turn-O-Graph, 11, 14, 15, 22, 91, 99-105, 129
    model 6202, **14**, 15, 71, 100, **100**, **101**, 102,
    102, 103, **103**, 104, **104**, 105, **105**, 129
typeface: serif, 82, **82**, 85
typeface: slab serif, 81, **81**, 85

waterproof, 10, 11, 14, 15, 20, 25, 49, 70, 90,
101
Wilsdorf, Hans, 8, 9, 12
winding crown, 22, 27, 30, 116
World War II, 25

Movement Index

72B movement; Valjoux, 108, 112, 112, 114,
129
A260 movement, 20, 21, 26, 27, 28, 29, 100,
102, 128, 129
A296 movement, 20, 21, 27, 30, 34, 71, 74, 75,
76, 100, 103, 104, 105, 128, 129
722/1 movement; Valjoux, 108, 113, 114, 129
727 movement; Valjoux, 109, 115, 117, 129
1000/419343 movement, 26, 128
1030 movement, 22, 30, 31, 21, 33, 71, 76, 77,
78, 128, 129
1036 movement, 58, 60, 128
1065 movement, 58, 61, 90, 128
1065M movement, 90, 92, 93, 129
1066 movement, 58, 61, 62, 65, 90, 128
1066M movement, 90, 93, 94, 129
1080 movement, 91, 95, 129
1520 movement, 23, 39, 42, 43, 44, 52, 53, 72,
80, 81, 83, 128, 129
1530 movement, 22, 23, 34, 35, 37-41, 71, 78-
80, 128, 129
1560 movement, 23, 36, 38, 41, 71, 82, 128,
129
1565 movement, 59, 63, 64, 128
1570 movement, 23, 36, 71, 73, 81-83, 85-87,
128, 129
1575 movement, 23, 42, 44, 49, 53-55, 59, 64-
66, 73, 84, 85, 128, 129
1580M movement, 96, 97, 129
3035 movement, 55, 128
3075 movement, 59, 67, 128
3085 movement, 45, 73, 86, 128, 129

Bibliography

Books

Aritake, Shigy. *Rolex Scene, London Aritake Collection 1913-1997*. Japan: World Mook, 1998.

Dowling, James and Jeffrey P. Hess, *Rolex wristwatches: an unauthorised History*. Atglen, Pennsylvania; USA: Schiffer Publishing Lt, 1996.

Ehrhardt, Sherry and Roy Ehrhardt. *Wrist watch price guide*. Kansas City, Missouri; USA: Heart of America Press, 1988.

Imai, Kesaharu. *Rolex: 2421 Uren*. Munich, Germany: Callwey, 1998.

Introna, Elena and Gabriele Ribolini. *Maitres Du Temps*. Paris, France: Du May, 1992.

Kahler, Helmut, Richard Muhe and Gisbert L. Brunner. *Wristwatches*. Atglen, Pennsylvania; USA: Schiffer Publishing Lt, 1986.

Lang, Lang R and Reinhard Meis. *Chronograph wristwatches*. Atglen, Pennsylvania; USA: Schiffer Publishing Lt, 1993.

Patrizzi, Osvaldo. *Collecting Rolex wristwatches*. Genova, Italy: Guido Mondani Editore, 1999.

Rolex Watch Company, Alfred Chapuis and Eugene Jaquet. *Rolex Jubilee vade Mecum*. Switzerland: Rolex Watch Company, 1946.

Viola, Gerald and Gisbert L. Brunner, *Time in Gold: wristwatches*. Atglen, Pennsylvania; USA: Schiffer Publishing Lt, 1998.

Periodicals

National Geographic, Washington D.C, USA: National Geographic Society, 1953 to 1990.

Christie's auction catalogues, Geneva, London and New York, 1990 to date.

Sotheby's auction catalogues, Geneva, London and New York, 1990 to date.

The Times Newspaper, 1953 to 1956.

Neptune, B.S.A.C Journal. London, England. 1954 to 1956.

Triton, B.S.A.C Journal. London, England: Eaton Publications, 1956 to 1958.